MAGNETIC IONS
IN CRYSTALS

Magnetic Ions in Crystals

K. W. H. STEVENS

PRINCETON UNIVERSITY PRESS

PRINCETON, NEW JERSEY

Library of Congress Cataloging-in-Publication Data

Stevens, K. W. H.
Magnetic ions in crystals / K.W.H. Stevens
p.   cm.
Includes bibliographical references and index.
ISBN 0-691-02693-9 (alk. paper). —
ISBN 0-691-02692-0 (pbk. : alk. paper)
1. Magnetic ions.   2. Crystal field theory.
3. Hamiltonian systems.   I. Title.
QC754.2.M333S74   1997
530.4'12—dc20   96-33485   CIP

This book has been composed in Times Roman using LaTeX

Printed in the United States of America by Princeton Academic Press

1   2   3   4   5   6   7   8   9   10

Contents

Preface

The strength of a chain is that of its weakest link.

It seems likely that it should be possible to explain the great diversity in properties shown by liquids and solids by applying nonrelativistic quantum mechanics to a Hamiltonian which is almost a universal operator, where "almost" can be removed if it is allowed to specify the number of electrons and the number and types of nuclei. So far, though, apart from a few standard examples used for teaching purposes, there seems to have been little progress in this direction, for there are few, if any, exact results for the eigenvalues and eigenfunctions of such a Hamiltonian when more than two particles are present.

The development of the technique of electron spin resonance, or paramagnetic resonance, as it is alternatively described, has indicated a way by which some progress may be possible, for it has shown that the low-lying levels of many magnetic impurities in nonmagnetic host crystals can be accurately represented by comparatively simple effective Hamiltonians, generally known as spin-Hamiltonians. It has thereby stimulated interest in the question of how their forms may be derived from a generic Hamiltonian.

This is not a new problem, for from about 1950 onwards there was a good deal of interest in using crystal field theory to do just this, and indeed it was via this theory that spin-Hamiltonians came into being. But what is surprising is that, almost from the beginning, it was known that Professor J. C. Slater had expressed concern about the validity of the whole procedure on the grounds that it is a theory that distinguishes the electrons on the central ion from those on the ligands and so violates the uncertainty principle, which is a fundamental principle of quantum mechanics. As time has passed it has been increasingly accepted that this is a major weakness of the theory and of most of its direct extensions. So while the title I have chosen for this book may suggest that it is about crystal field theory this is incorrect, for its purpose is to provide an introduction to an alternative method, one that is relatively undeveloped, which does not violate the exclusion principle as applied to electrons and which is based on a generic Hamiltonian. It will be shown that it produces the spin-Hamiltonians for isolated magnetic ions just as readily as do the theories based on crystal field concepts. It will also be shown that it can be extended into areas where crystal field theory runs into even more difficulty.

My first experience with crystal field theory came in the context of a study, by electron spin resonance, of the nickel ion in nickel fluorosilicate, a crystal in which each Ni^{2+} ion is at the center of a trigonally distorted octahedron of identical nonmagnetic neighbors. Having worked through the theory I arrived at an effective Hamiltonian:

$$-\delta S_z^2 - \beta H_z g_\| S_z - \beta g_\perp (H_x S_x + H_y S_y),$$

where S_x, S_y, and S_z are the three components of an effective spin, $\mathbf{S} = \mathbf{1}$, though as the Ni^{2+} also has an actual spin of $\mathbf{S} = \mathbf{1}$ it was readily possible to confuse the two. An extensive range of experimental observations, at 90 K, could be fitted by choosing $\delta_\| = 0.17\,\mathrm{cm}^{-1}$ and $g_\| = g_\perp = 2.29$, values which it was possible to account for by making what seemed reasonable assumptions about the magnitudes of the octahedral and trigonal components of the crystal field.

This effective Hamiltonian is an example of a spin-Hamiltonian. It is a particularly simple one, for as the technique of electron paramagnetic resonance has advanced so more complicated forms have been introduced. Many examples can be found, for example, in a book by Abragam and Bleaney (1970), which will henceforth be denoted by (A-B). There are usually more parameters and more elaborate spin operators than those in the above example and its properties have usually been confirmed to a much higher degree of accuracy.

Several inferences can be drawn. If crystal field theory is to be replaced by another theory, it can be expected that the new theory should lead to spin-Hamiltonians of the same forms as those already in use, however they have been obtained. (A number seem to have been obtained by inspired guesswork!) Another is that the expressions for the parameters, such as δ and g, are likely to differ from those emerging from a theory that violates the exclusion principle, so any agreement between observation and theory, based on such a theory, can only be regarded as fortuitous.

In the light of the above remarks it therefore seems relevant to delineate some of the features that will underlie most of what is to follow. The first is a belief that it is possible to write down a generic Hamiltonian which will be such that, although it will only be studied here in the context of specific magnetic systems, it is much more general, for it contains all that is needed for describing most of the properties of the many-particle systems with which we are surrounded; that is, of systems which exist at, say, temperatures below 1000 K. (The reservation implied by "most" has been made because all the systems in mind are subject to radiation of one sort or another from a variety of sources at higher temperatures. Nor are they completely isolated from similar systems.) The main terms in the generic Hamiltonian will be those that describe the kinetic energies of the particles and those that describe their mutual electrostatic potential energies. All that is necessary is to know the number of electrons and the number and nature of the various nuclei. This part of the generic Hamiltonian can be written down by analogy with the classical Hamiltonian of such a set of particles.

The position is less clear for the smaller terms, some of which are known to contribute significantly to magnetic effects. The problem is basically that the most reliable information about them has come from systems containing just one nucleus, atomic hydrogen, and the many-electron atoms and ions. These studies have led to the addition of extra terms to the generic Hamiltonian, modifications that seem essential to bring theory and experiment into agreement. When two or more nuclei are present, the experimental information is usually subject to a wider margin of error and the modifications needed to bring agreement between theory and experiment are then more difficult to formulate reliably. It may therefore be an important feature of spin-Hamiltonians that not only do they provide information about systems containing a number of nuclei but they also throw light on some of the smaller terms in the Hamiltonian.

Setting up the machinery of the new method is quite a lengthy process. It will be assumed that the reader is familiar with the standard techniques of quantum mechanics but will nevertheless welcome a number of introductory chapters on perhaps less familiar material, as preliminary steps for what is a major target, an attempt to use what is known about magnetic ions in crystals to set up a perturbation procedure that will account for their properties, beginning with the generic Hamiltonian. On the way it will be necessary to introduce criteria that pinpoint the properties of magnetic ions that allow the treatment to proceed. That they are not always essential is later demonstrated, by showing that the initial approach, which relies on the insulating nature of the crystals, can be modified to allow an extension of the method to some types of magnetic conductors.

The main thrust of the program can be regarded as an attempt to meet a challenge. Is it possible to begin with the generic Hamiltonian of a nonmagnetic crystal containing one magnetic ion, a description that fixes the number of electrons and nuclei, and find a viable path, one which does not violate any fundamental principles and is mathematically sound, which shows that its low-lying energy levels can be described by an effective operator of spin-Hamiltonian form?

Turning now to the various chapters, the first is mainly concerned with the question of how the form of the generic Hamiltonian has been established. Then, as experience has shown that the symmetry properties of Hamiltonians can be a considerable help in the examination of their properties, the second chapter is devoted to those parts of group theory that are known to be particularly valuable. This is followed by a chapter on perturbation methods, which gradually focuses on the method that is basic to the rest of the book. A fairly standard account of crystal field theory is then given, mainly because it provides a number of excellent examples of how group theory and perturbation theory can be advantageously combined, but also because it shows how the spin-Hamiltonians have usually been derived, so establishing a target for the new theory.

Chapter 5 is somewhat different, for it is aimed at preparing the way for introducing the new way of looking at magnetic ions in crystals. It is therefore unlikely to be familiar to those who have been brought up on crystal field theory. The next chapter gets down to doing this in earnest, by the introduction of the technique of second quantization, which is in almost universal use in the rest of solid state physics but which has been almost entirely neglected in the context of magnetic ions. Chapter 7 is then devoted to showing how, in the appropriate systems, a combination of the formalism of second quantization, a form of perturbation theory due to Bloch, and a knowledge of group theory can be used to obtain spin-Hamiltonians of the same forms as have previously been obtained using crystal field theory. Chapter 8 is a straightforward extension of the methods of Chapter 7 to pairs of magnetic ions and so, by extrapolation, to assemblies of magnetic ions.

Chapter 9 marks another change in direction, one that is needed when there is no experimental technique that can be used to determine the detailed pattern of energy levels. There is then no way of determining, with complete confidence, the parameters in the spin-Hamiltonians, even though the theory gives their forms. As the interacting ions show new phenomena, such as phase changes as the temperature is lowered, from paramagnetic to cooperative phases, and spin-wave spectra at even lower temperatures, the interest is in showing that the theoretical spin-Hamiltonians have the potential for describing these. This chapter is therefore mainly about the problems associated with phase changes and spin-wave theory.

Chapter 10 is again different, and it is likely to offer a substantial challenge to the reader, for it introduces a number of new considerations associated with the possibility that magnetic ions may have variable valencies, which will allow conduction to occur. The method used in the previous chapters has, therefore, to be extended to allow for the lattice distortions that accompany the "hopping" of electrons from one site to another. The theory is found to have features so similar to some of those used in the BCS theory of superconductivity that it has seemed worthwhile to pursue the analysis to some depth.

Finally comes Chapter 11. I hesitated over including this because the reader may well conclude that it undermines much of what has gone before, where care has been taken not to distinguish between electrons. The issue is that, along with a good many other theoreticians, I have not taken the same care with identical nuclei. Being aware that this is a potential criticism, it is one which I have felt I should not ignore. I have therefore attempted to show that nuclear symmetry can probably be safely ignored in most, but not all, cases and that the latter merit further study.

Acknowledgments

This book, which was begun during the tenure of an Emeritus Fellowship awarded by the Leverhulme Trust, has been written in an office provided by the Department of Physics at the University of Nottingham, where I have enjoyed the stimulus and company of a congenial group of talented physicists. I would therefore like to take this opportunity to express my appreciation and thanks to the Trust, to my Nottingham colleagues, and to all the other friends I have made through physics.

References

Abragam, A., and Bleaney, B. 1970. *Electron Paramagnetic Resonance of Transition Ions* (Clarendon Press: Oxford), referred to as (A-B).

MAGNETIC IONS IN CRYSTALS

1

Introduction

§ 1.1 Microscopic Magnetism

From a microscopic point of view magnetic properties arise from two sources, the translational motion of charged particles, particularly electrons, and the magnetic moments associated with the spins of electrons and nuclei. The steps that eventually led to this picture began in the early 1900s, when Zeeman studied the changes that occur in the optical spectra of free ions in the presence of external magnetic fields. It was still a long road though, through classical physics, Bohr theory, quantum theory, and the introduction of a wide range of experimental techniques, before the present understanding of the properties of magnetic ions was established. It was initially assumed that spectral lines were in some way associated with harmonic oscillations, but in spite of a good deal of analysis and ingenuity little progress was made until the Bohr theory came along, in 1915. This substituted transitions between energy levels for harmonic-oscillator-like resonances. Rapid progress then ensued, at least in interpreting the spectroscopy of the simplest ions, but gradually this too slowed down as it became increasingly difficult to explain the observations on more complicated ions [see Van Vleck (1926) for a detailed account of the position just before the advent of quantum mechanics]. Quantum mechanics, in the late 1920s, opened a new avenue, one which was quickly recognized as being more promising and which could be rapidly developed in its wave-mechanical form using mathematical techniques familiar from the Bohr theory. Within just a few years, enough progress had been made for Condon and Shortley to begin writing, in 1931, a detailed account of the theory of atomic spectra. The book, which appeared in 1935 (with a later revision, Condon and Shortley, 1967), is still one of the standard texts on atomic spectroscopy.

§ 1.2 Ions in Solids

With the theory of the isolated ions in a satisfactory form, attention turned to using it to explain the spectroscopic properties of molecules and the macroscopic properties of liquids and solids. In the context of magnetism, it was known that iron was a ferromagnet, that many metals had temperature-independent paramagnetic susceptibilities, that many ionic crystals were diamagnetic, and that others had susceptibilities that obeyed either the Curie or the Curie-Weiss law, $\chi = C/(T + \Delta)$, where C and Δ ($= 0$ in the Curie law) are constants and T is the absolute temperature. (In paramagnets the moment is in the direction of the applied field, while in diamagnets it is in the opposite direction. For each, the susceptibility is defined as the ratio of the magnetic moment to the magnetic field in the limit when the field tends to zero.) The paramagnetism of the normal metals was fairly soon explained (in the 1930s) with the introduction of a simple form of band theory, but the ferromagnetism of iron, cobalt, and nickel remained a problem for much longer.

For the ionic magnetic crystals the direction to go had already been indicated, for even before the advent of the Bohr theory an explanation of their magnetic properties had been given. This assumed that the ions contained microscopic circulating currents, each of which would give rise to a magnetic moment. Thus in some crystals there might be "permanent" microscopic magnetic moments which in zero magnetic field would be randomly arranged and so cancel, whereas in others the internal moments would be such that there were no permanent moments to be canceled. An external magnetic field could be expected to induce a precessional motion of all the currents, which would produce a diamagnetic moment in both types of ion. However, in the ions with permanent moments it would also induce a partial alignment, and it was the thermal distribution over the associated energies that added a temperature-dependent paramagnetism. [Although the theory was worked out in some detail and seemed to be satisfactory it was later realized that it was unsound. See Chapters 4 and 5 of Van Vleck (1932), which, incidentally, seems to have been the first book to apply quantum mechanics to solid state problems.]

The advent of the quantum theories led to the concept of electron shells, and when this was supplemented by the idea that no two electrons could have the same quantum numbers (originally three in number, but increased to four when electron spin was discovered), the theory led to an understanding of the magnetic properties of the free atoms and ions and an understanding of the periodic table of the elements. In the lowest-energy arrangement, the shells are filled progressively in energy so that in some atoms and ions there are just the right number of electrons to fill all the occupied shells. Between these, there will be others in which the outer shells, usually just one, are partially filled. Those atoms or ions with filled shells have no orbital and no spin moment so they can be identified with the entities which give diamagnetism. Those with partially filled outer shells usually have all the inner shells filled (the rare earths are an exception), so diamagnetism can be associated with their inner shells and paramagnetism with their partially filled outer shells. The chemical properties follow when the free atom picture is supplemented by the idea that in compounds electrons have moved from one atom to another to produce ions with closed-shell structures.

It was also known that in many of the iron and rare earth group compounds the ions do not have closed-shell structures, and it was therefore not surprising that they showed properties typical of those expected from microscopic permanent magnetic moments. There was, however, a problem, particularly for the iron group compounds, that the moments of the ions in a crystal, obtained from susceptibility measurements, were usually quite different from those of the same ion in free space, obtained from the Zeeman effect. (Less was known about the susceptibilities of rare earth compounds. Over the temperature ranges that had been used the agreement seemed much better. This did not last when measurements were made at lower temperatures.)

The step which led to the resolution of the problem was taken in the early 1930s, by Van Vleck, who introduced what is known as the crystal field. Each magnetic ion was regarded as being subject to an electric field due to its ionic nonmagnetic neighbors, and the theoretical problem became one of including this interaction in the Hamiltonian of the free ion, which already included a number of interactions of differing magnitudes and which had been treated by a sequence of approximations. An important consideration was that large perturbations needed to be treated before smaller ones, so it was necessary to estimate the magnitude of the crystal field perturbation, incorporate it in the right place in the perturbation sequence, and work out the consequences. It is not necessary to go into any details at this stage, but two remarks can be made. The first is that although the theory is an extension of the isolated ion theory, much of the detail of that theory is not needed, for magnetism is a property which, in the context of solids, is confined to a restricted thermal range. So in the theory the only energy levels of interest are those in a range which extends to not much more than kT above the ground level, where k is Boltzmann's constant and T is the absolute temperature. The other remark is that the theory appeared to be very successful.

In the late 1940s the technique of electron paramagnetic resonance (EPR) was introduced. In due course it was used to study the low-lying energy levels of isolated magnetic ions, impurities in an otherwise nonmagnetic host. (Nonmagnetic ions are ions with closed-shell structures. As they typically show a small temperature-independent diamagnetism they are not, strictly, nonmagnetic. Nevertheless, they are so described to distinguish them from the ions that have partially filled shells of electrons, which typically show much larger paramagnetic susceptibilities.) Until then, crystal field theory had been used to explain thermal properties, which invariably involved taking averages over energy level distributions and examining how they change under the influence of external perturbations, such as the application of a magnetic field. The spectroscopic nature of EPR gives energy differences, so thermal considerations enter only in the intensities of the resonance lines. EPR (which tends to be used interchangeably with ESR), therefore provided a much more sensitive test of the validity of crystal field theory.

The results are almost always expressed using the concept of spin-Hamiltonians, which is apt to give the impression that they depend on crystal field theory. In fact some of them, such as those for covalently bonded ions, have come from a theory which is much more like that used for molecules, and for a few others, particularly S-state ions, it is not obvious where they have come from.

Nevertheless, in the early days of EPR it seemed that it was the crystal field that provided the explanation of the observations, and even now the way of describing the underlying theory usually involves a reference to crystal fields, with, perhaps, a few remarks that modifications may be needed (e.g., delocalization of orbitals and/or covalency). It is, therefore, of interest to recall that at about the same time that crystal field theory was being actively developed, band theory was coming into prominence as the way in which to describe periodic systems. It would have been possible to claim that there was no conflict between the two because there was no common ground, for crystal field theory was being verified only for isolated magnetic ions in otherwise diamagnetic crystals. This would clearly have been unconvincing because there was no doubt that it was expected that it could be applied to fully concentrated, and therefore periodic, magnetic crystals (with a few additional interactions such as magnetic dipole and exchange interactions between the ions). The exponents of band theory believed that their way of proceeding would eventually lead to a better explanation of the resonance results, and that it was incorrect to use a model in which the electrons are allocated to specific sites rather than placed in delocalized orbitals that extend throughout the crystal. The weakness of this argument was the lack of evidence that it could come anywhere near crystal field theory in explaining the resonance results.

The second criticism was that the electrons on the magnetic ions were being distinguished from the electrons on the neighboring ions, which were simply regarded as producing a large part of the crystal field. A fundamental principle, that electrons are indistinguishable, was therefore being violated.

Of these two criticisms, it is now established that the first is not a flaw. The Hamiltonian of a periodic arrangement of ions is a complicated expression, and it is certainly an approximation to replace the detailed interactions between a large number of charges by an assumption that each electron moves in a periodic potential. Doing so inevitably leads to a description in terms of bands. But the approximation is not necessarily a good one. The second criticism is one that needs to be taken seriously, though it was not until the 1970s that a satisfactory way out of the difficulty was found.

An alternative to crystal field theory was proposed by Van Vleck (1935) at almost the same time that he was putting forward the concept of a crystal field. It was based on the idea of regarding the magnetic ion and its neighbors as a unit in which the electrons move in orbitals that extend over the whole complex. It is therefore rather similar to the picture used in band theory, the main difference being that the effective potential is not periodic. It has been extensively developed, and is usually referred to as ligand field theory. It runs into much the same difficulty as occurs in band theory, the step of replacing energies in a fictitious electrostatic field by the actual Coulomb interactions. There are a number of other theories, an example being the superposition model, which endeavors to relate the spin-Hamiltonian parameters for an ion with many neighbors to those that would be found for the same ion with just one neighbor. (As the purpose of this book is to describe a way in which the problems with these theories can be circumvented, there seems little need to describe them in detail here. They do, though, form an important part of the development of the understanding of electron spin resonance. It has therefore seemed only right to give the reader an entry into what is an extensive literature, through a selection of references which should provide an entry into the whole field. These can be found at the end of this chapter.)

The new method relies a good deal on symmetry arguments and perturbation theory, as does crystal field theory, so instead of going immediately to its description it has seemed better to introduce it through crystal field theory. It then becomes possible to make comparisons of the two methods which, it is hoped, will reveal the rather subtle differences between the two and show the logical advantages of the new method. As there are a number of theoretical points that form a common background to both theories, the rest of this chapter will be devoted to a discussion of these.

§ 1.3 The Choice of Hamiltonian

Until early in the twentieth century, dynamics was based entirely on Newtonian mechanics, and its inadequacies only became apparent when particles of atomic dimensions came to be studied. Two revolutionary changes then occurred, the discoveries of quantum mechanics and relativity. For most solid state purposes, relativistic considerations are of minor importance and most theoretical work therefore uses nonrelativistic quantum mechanics. This practice will be followed here. It produces many simplifications while still leaving a number of questions open, one being that of how to choose the generic Hamiltonian. A reader coming new to solid state theory might suppose that there is no such problem, because Hamiltonians are frequently written down with no justifications except, possibly, that they have been used in previous work. Turning to the elementary texts on quantum mechanics for guidance is unlikely to be much help, for they are usually concerned with the exact mathematical treatments of a relatively small number of model Hamiltonians, which have been obtained from classical mechanics, with no mention of their relevance to solid state physics.

From one point of view this is surprising, for quantum theory might be expected to be quite different from classical mechanics. But from the purely practical point of view one can see why it was done initially, for what other choice was there? Elements of this approach still remain, but so much has now been learnt that it is well established that while the main "classical energies" of a solid are indeed present, as operators, there should also be a number of other operators, which have been shown to be present in the Hamiltonians of isolated atoms and ions. There is also the nonclassical requirement that the wave functions of many-electron systems must be antisymmetric with respect to interchanges of electrons.

It is therefore possible to begin writing down a generic Hamiltonian which should be adequate for all solids. Indeed, for many purposes it is probably sufficient to write down just the main terms, assuming the nuclei are fixed, which are

$$\sum_i \frac{\mathbf{p}_i^2}{2m} \tag{1.1}$$

to represent the kinetic energy of the electrons, i being a label that runs over all the electrons,

$$-\sum_{i,I} \frac{Z_I e^2}{|\mathbf{r}_i - \mathbf{R}_I|}, \tag{1.2}$$

which represents their Coulomb energy in the field of the nuclei, I being a label for nuclei, and

$$\frac{1}{2} \sum_{i \neq j} \frac{e^2}{r_{ij}}, \tag{1.3}$$

which represents the energy associated with the mutual Coulomb repulsion between the electrons. If the nuclei are not fixed there should be extra terms to represent their kinetic energies and their mutual electrostatic energies.

Unfortunately, no exact eigenvalues, eigenstates, and solutions for the Schrödinger equation are known for any Hamiltonian of the above form for a system that contains more than two particles. This is probably why it is not usually stressed that the great diversity of properties shown by solids all result from what is basically the same Hamiltonian, it being necessary only to vary the number of nuclei, their charges, and the number of electrons.

The diversity is more commonly accounted for by the variety of approximations made in dealing with the generic Hamiltonian, which is one of the reasons why only the above energies have been given specifically, for there is usually enough difficulty in dealing with these without including any more, particularly as there is no virtue in making approximations unless progress can then be made. This is where the standard problems of the textbooks come in useful, for a common aim is to approximate so as to reduce the problem to one that is standard and soluble.

The usual procedure will not be followed in the method to be described, for the primary aim is to show how the generic Hamiltonian for an isolated magnetic ion in a crystal can be reduced to a spin-Hamiltonian form. It is therefore more ambitious in one sense and less ambitious in another. The insistence on using the generic Hamiltonian is the ambitious feature, and the less ambitious one is that of showing how the known forms of the spin-Hamiltonians of isolated ions emerge. Even for isolated ions there can be minor problems, usually resolvable with computers, in finding the eigenvalues and eigenfunctions. For fully concentrated systems the problems appear, at present, to be quite intractable.

There is one standard problem that is of fundamental importance because its predictions can be tested directly by experiment. This is the hydrogenic problem: one electron and one nucleus. With just the kinetic and electrostatic energies it is completely soluble and the comparison has shown that such a Hamiltonian is insufficient. Extra nonclassical terms have therefore been added to the Hamiltonian, and the need for them has indicated that similar terms should also be added to the generic many-particle Hamiltonian. Thus, in developing the theory of the magnetic ions, there will need to be an acceptance that at some time the generic Hamiltonian may need to be supplemented with extra interactions.

For the next section this need not be a concern, for its main purpose is to use hydrogen to illustrate some general properties of angular momentum in quantum mechanics.

§ 1.4 The Hydrogenic Atom

The basic problem is treated in many standard texts, so only an outline of some of the results will be given. It is commonly shown that by an appropriate change of variables the Hamiltonian can be separated into the motion of the center of mass and motion relative it. The center of mass motion is then that of a free particle, and the internal motion is that of an electron with a slightly changed mass in a potential that varies inversely as its distance from the origin. The Hamiltonian for the internal motion can be taken as

$$H = \frac{\mathbf{p}^2}{2m} - \frac{e^2}{r}. \tag{1.4}$$

The momentum variables, when expressed in wave-mechanical form, are

$$p_x = -i\hbar \frac{\partial}{\partial x}, \; p_y = -i\hbar \frac{\partial}{\partial y}, \; p_z = -i\hbar \frac{\partial}{\partial z}, \tag{1.5}$$

and the kinetic energy part becomes a multiple of the differential operator:

$$\nabla^2 = \frac{\partial^2}{\partial x^2} + \frac{\partial^2}{\partial y^2} + \frac{\partial^2}{\partial z^2}, \tag{1.6}$$

which is invariant under any rotation of the Cartesian axes, as is the term describing the Coulomb interaction. The Hamiltonian, when written in spherical polar coordinates (r, θ, ϕ), separates, and each wave function becomes a product of a radial function and an angular part. The radial parts of the wave functions will not be given explicitly because, unlike the angular parts, they are of little interest in the context of many-electron ions in solids. The particular interest of the angular parts is that they are the same as those of the eigenfunctions of the operator in (1.6) when this is written in polar coordinates. When appropriately defined they are standard mathematical functions.

Most textbooks give the orthonormal eigenfunctions of hydrogen in the form

$$\Psi(r, \theta, \phi) = \frac{R(n, l)}{r} Y_l^m(\theta, \phi), \tag{1.7}$$

where $R(n, l)/r$ is the radial part of the solution and the rest is the angular part of an eigenfunction of (1.6). n and l take positive integer values (including zero for l) with $n > l$ and m taking positive and negative integer values between l and $-l$. θ lies between 0 and π and ϕ lies between 0 and 2π. [A word of caution is necessary, for the definitions of the Y_l^m vary between authors (Killingbeck and Cole, 1971).] Here they will be taken as

$$
\begin{aligned}
Y_l^m(\theta, \phi) &= (-1)^{(m+|m|)/2} \left[\frac{(2l+1)(l-|m|)!}{2(l+|m|)!} \right]^{1/2} \\
&\quad \times \frac{1}{(2\pi)^{1/2}} P_l^{|m|}(\theta) \exp(im\phi),
\end{aligned}
\tag{1.8}
$$

where P_l^m is another standard function, an associated Legendre function. The eigenvalues for hydrogen are found to depend only on n, so the presence of the l and m quantum numbers in the eigenfunctions indicates that most energy levels are degenerate. (A level is degenerate if there are a number of orthogonal eigenfunctions with the same energy. Many sets of mutually orthonormal eigenfunctions can be formed from them, without changing the number of eigenfunctions in each set. The number defines the degeneracy.)

The amplitudes and phases chosen for the Y_l^m are such that the various states can be obtained from one another by the application of raising and lowering operators, which are defined as

$$l_+ = l_x + il_y, \, l_- = l_x - il_y,$$

where $\mathbf{l} = (\mathbf{r} \wedge \mathbf{p})\hbar$. By using the commutation rule $\mathbf{l} \wedge \mathbf{l} = i\mathbf{l}$ it can be shown that when the wave-mechanical state defined in eqn. (1.7) is put into correspondence with a ket vector $|n, l, m\rangle$,

$$l_+|n, l, m\rangle = [l(l + 1) - m(m + 1)]^{1/2}|n, l, m + 1\rangle,$$

and

$$l_-|n, l, m\rangle = [l(l + 1) - m(m - 1)]^{1/2}|n, l, m - 1\rangle.$$

(These are relations in quantum mechanics. They are consistent with the wave-mechanical relations obtained by writing the raising and lowering operators in differential form and replacing the kets by the wave functions.) It should be noted that although the l_x, l_y, and l_z operators are usually referred to as angular momentum operators, they need to be multiplied by $\hbar$ to have the correct dimensions. The radial parts of the wave functions do not depend on m.

In deriving the eigenfunctions, and particularly if the angular parts have been found from (1.6), there is an assumption that at each point in space each is uniquely determined. So each eigenfunction must be unaltered if θ and ϕ are increased by 2π. In fact, the physical requirement is that it is the square modulus of the eigenfunction which should be unique, which means that there is no direct requirement on the eigenfunction itself. However, there is also a requirement that the eigenfunction can be normalized. This leads to a requirement that its radial part must fall to zero as $r \to \infty$, and it is this condition that leads to the 2π angular condition. The energies associated with these eigenfunctions are negative and given by the expression $E = -(me^4)/(4\pi c\hbar^3 n^2)$. (They are called internal energies because there are other eigenfunctions that have positive energies. They are associated with the decomposition of the atom into a bare proton and an unbound electron, states with a continuum of positive energies.)

It is readily seen that the Hamiltonian has spherical symmetry, so it is consistent with the physical expectation that the properties of hydrogen should be independent of the choice of orientation of the x, y, z axes. This is an example of a widely expected symmetry of quantum systems, and here it is particularly interesting because it can be examined in the context of a soluble problem, for any property due to symmetry can be expected to carry over to insoluble problems which have the same symmetry. For $l = 0$ the Y_l^m also has $m = 0$ and so is a constant. Thus all eigenfunctions with $l = 0$ depend solely on r and are rotationally invariant. For the other values of l this is not so, for the individual eigenfunctions are not rotationally invariant. The shape of each eigenfunction with $l > 0$ and a specific m depends on the directions chosen for the axes from which θ and ϕ are measured. If another set is chosen, the corresponding eigenfunction (same n, l, and m) will have the same radial function but a different angular dependence when viewed relative to the first choice of axes. It is, however, a property of the associated Legendre functions that, for a common value of l any member of either set can be expressed as a linear combination of the members of the other set. It follows that the eigenfunctions of a given n

and l for one choice of axes are no more special than the choice for any rotated set. This is the sense in which the rotational symmetry is preserved.

The eigenfunctions with $m \neq 0$ are complex. Those with nonzero m and $-m$ can be combined to give two real functions, which is useful in trying to picture them. For $l = 1$ the three real functions, the two formed with $m = 1$ and $m = -1$ and the $m = 0$ function, have the same angular dependences as x, y, and z. It is then obvious that not only is each function nonspherical, but each is related to the others by a rotation. If, further, a different choice of axes is made and the corresponding functions are labeled X, Y, and Z then it is immediately clear that any one of them can be written as a linear combination of x, y, and z.

For most algebraic purposes it is better to replace wave mechanics by quantum mechanics and perform as many manipulations as possible by using the properties of operators. The papers which have definitions of the Y_l^m that are inconsistent with the angular momentum rules in wave mechanics invariably use the correct commutation rules for the angular momentum operators, with the result that underlying inconsistencies in the definitions tend to be unimportant as well as unnoticed.

For $l = 2$ there are five eigenfunctions for each value of n. They can be combined to give five real functions which have the same angular dependence as xy, yz, zx, and $x^2 - y^2$, $3z^2 - r^2$. (These are not orthonormal forms; numerical multiples have been omitted for convenience.) In setting them out they have been separated into two sets, for by inspection it is readily seen that the first three have the same shapes, differing only in their orientations in space. It is, therefore, not surprising that eigenstates with the same radial wave function and these shapes have the same energy. The fourth function also has the same shape, for apart from a numerical factor it corresponds to xy when this has been rotated by $\pi/4$ about Oz. It is $3z^2 - r^2$ which has a shape that is really different. It might, of course, have been expected that with $x^2 - y^2$ in the set there would also be $y^2 - z^2$ and $z^2 - x^2$. These three functions sum to zero, so any one of the three can be written as a linear combination of the other two. Thus, while the first four in the set are mutually orthogonal (no one of the set can be expressed as a linear combination of the other three), it requires one more function to complete the set. The usual choice is $3z^2 - r^2$, where $r^2 = x^2 + y^2 + z^2$. (The demonstration that $y^2 - z^2$ and $z^2 - x^2$ can be written as linear combinations of $x^2 - y^2$ and $3z^2 - r^2$ is left to the reader as an exercise.)

For historical reasons the eigenfunctions with $l = 0$ are referred to as s states, those with $l = 1$ as p states, those with $l = 2$ as d states, and those with $l = 3$ as f states. (The higher l values seldom occur in solid state theory, but where they do the notation continues in an alphabetic sequence, from g onwards.)

Each eigenfunction of the Hamiltonian is also an eigenfunction, with eigenvalue $l(l + 1)$, of $\mathbf{l}^2$ or $l_x^2 + l_y^2 + l_z^2$, the operator which represents the square of the total angular momentum in units of $\hbar^2$. When it is expressed in the form given in eqn. (1.8) or as $|n, l, m\rangle$ it is also an eigenfunctions of l_z, with eigenvalue m. The solution of lowest energy is obtained with $n = 1$. As it has $l = 0$ and so is of s type, it is often denoted $(1s)$. The next highest energy level corresponds to having $n = 2$ in the expression for E. It has the rather surprising feature that there is one s family and one p family with this eigenvalue. They are denoted $(2s)$ and $(2p)$, respectively. The next level, $n = 3$, has coincident s, p, and d families, which are denoted $(3s)$, $(3p)$, and $(3d)$, respectively, and so on for all the higher n values.

A degeneracy, a number of orthogonal eigenfunctions having the same eigenvalue, is usually associated with symmetry properties of the operator that has the eigenvalue. The m degeneracy associated with each l value is due to the spherical symmetry of the Hamiltonian. It is less evident that the degeneracy of a number of different l values with the same n is also due to a symmetry property of the Hamiltonian. This has been investigated and it has been shown that it is associated with a property of $1/r$, for with some other dependence on r, say $1/r^{1.5}$, the l degeneracy disappears. There is no need to pursue this in detail here, except to give a source of references (Thompson, 1994) and note that it is not always easy to spot symmetry properties of operators. So if an "accidental degeneracy" is found, it may be worthwhile seeking a symmetry reason for it.

Another type of "degeneracy" occurs with the Hamiltonians of magnetic systems. The Hamiltonian contains terms that depend on the direction and magnitude of the applied magnetic field, both of which can be varied, so the energy levels change in a continuous way. It can then happen that it is predicted that at certain field values and directions two energy levels will coincide, so creating a "degeneracy." In fact, there seldom is a degeneracy, for while it has been convenient to describe how it might occur, it is not what happens in practice. It is usually an artifact of the Hamiltonian. An approximate Hamiltonian has been used, obtained by dropping terms that are thought to be small. When these are restored it is usually found that the energy level pattern is particularly sensitive to them in the vicinity of the "crossing," with the result that there is no crossing. Instead, the energy levels appear to repel one another.

Families of states which are degenerate because of a symmetry property are generally of particular interest in spectroscopy, for there is always the expectation that by some appropriate technique the degeneracy can be lifted, in which case an observation of the splitting will establish that there really was a degeneracy. A favorite technique is to apply a uniform magnetic field, because it defines a direction and so reduces the full rotational symmetry to that of axial symmetry. This is why the Zeeman and the Paschen-Bach effects have been of interest. The former shows the lifting of a degeneracy which is linear in field strength and the latter shows that if, in consequence, two levels look destined to cross, there is actually a repulsion.

§ 1.5 The Concept of Electron Spin

As the accuracy of the determination of the spectrum of hydrogen increased it was found that the Hamiltonian had to be corrected, by the addition of terms which have no counterpart in classical physics. The most important of these are connected with the electron spin. This concept, which was introduced empirically by Uhlenbeck and Goudsmit (1925) in order to fit the experimental observations, led to the electron being regarded as spinning with an angular momentum, usually denoted by **s**, which is analogous to the **l** of the orbital moment in the commutation rules of its components, except that s is restricted to the value $1/2$ (Pauli, 1927). The theory of electron spin was later put on a firm theoretical foundation by the relativistic theory of Dirac (1947). The result was a doubling of the dimensions of the Hilbert space by the introduction of bra and ket vectors to describe the spin components. These are usually written as s_x, s_y, and s_z. They satisfy the commutation rule $\mathbf{s} \wedge \mathbf{s} = i\mathbf{s}$, with $s_x^2 + s_y^2 + s_z^2 = 3/4$. The associated kets, $|+\rangle$ and $|-\rangle$, are usually regarded as eigenstates of s_z which satisfy

$$s_z \mid +\rangle = \frac{1}{2} \mid +\rangle \text{ and } s_z \mid -\rangle = -\frac{1}{2} \mid -\rangle.$$

Instead of using s_x and s_y it is usually more convenient to use $s_+ = s_x + i s_y$ and $s_- = s_x - i s_y$, the analogues of l_+ and l_-. Then

$$s_- \mid +\rangle = \mid -\rangle, \ s_+ \mid -\rangle = \mid +\rangle, \text{ and } s_- \mid -\rangle = s_+ \mid +\rangle = 0.$$

The spin kets have an unusual property. In physical space the operator $\exp(i\alpha l_z)$ applied to an eigenstate of l_z with eigenvalue m multiplies it by $\exp(i\alpha m)$. It can be regarded as an operator which rotates any orbital state through an angle α about Oz, for it replaces ϕ in (1.8) by $(\phi + \alpha)$. With an integer value for m a rotation through 2π changes ϕ by $2m\pi$ and so brings any state into itself, as required by the boundary conditions. The corresponding operator in the spin space, $\exp(i\alpha s_z)$, reverses the sign of each spin ket under a 2π rotation and the initial state is only regained after a rotation through 4π!

A magnetic moment is associated with the spin of the electron, and the triumph of the Dirac theory was that not only did it explain the origin of the spin but it also gave its magnetic moment and an expression for the interaction of the spin with the orbital motion of the electron, the spin-orbit interaction. In its nonrelativistic form the spin-orbit interaction takes the form $\zeta \mathbf{l} \cdot \mathbf{s}$, where

$$\zeta = \frac{\hbar^2 e^2}{2m^2 e^2 r^3} \tag{1.9}$$

(Condon and Shortley, 1967). A new angular momentum operator, $\hbar \mathbf{j}$, where $\mathbf{j} = \mathbf{l} + \mathbf{s}$, can be introduced, and since all components of **s** commute with all components of **l** it follows that **j** has the same commutation properties as **l** and **s**, with the magnitude of j in

$$j_x^2 + j_y^2 + j_z^2 = j(j+1)$$

taking values equal to an integer plus one-half. j can thus be either $l + 1/2$ or $l - 1/2$, except when $l = 0$, when the only possible value is $j = 1/2$. j is therefore always positive. Also, because

$$2\mathbf{l} \cdot \mathbf{s} = \mathbf{j}^2 - \mathbf{l}^2 - \mathbf{s}^2 = j(j+1) - l(l+1) - s(s+1),$$

it follows that, for nonzero l, the effect of the spin-orbit interaction is to split what was previously a single energy level characterized by n, l into two levels characterized by n, j. The lower of these has $j = l-1/2$ and degeneracy $2j+1$ and the upper has $j = l + 1/2$, with degeneracy $2j + 1$, separated by ζ. Such a splitting had been observed even before the advent of quantum mechanics and attributed to a spin-orbit interaction. It even led to the introduction of a fourth quantum number (see Van Vleck, 1926). [In quantum mechanics the inclusion of spin doubles the degeneracy associated with l and produces a splitting, without changing the overall number, $2(2l+1)$, of states. One j level has $2l$ states and the other has $(2l + 2)$. This behavior is typical of the effect of introducing an additional interaction; while it may resolve a degeneracy, either partially or completely, it does not alter the overall number of eigenstates.]

§ 1.6 The Application of a Magnetic Field

The application of a uniform magnetic field produces a splitting of the degeneracy of a j manifold, where "manifold" has been introduced to indicate that the energy level associated with j is degenerate, which is such that the separations between adjacent levels are equal and proportional to the strength of the magnetic field. To explain this requires a modification of the Hamiltonian, to describe the effect of the field on the orbital motion of the electron and on the spin magnetic moment.

In classical physics a magnetic field changes the motion of a charge by producing a force that depends on the magnitude and direction of its velocity and is in a direction perpendicular to both the field and the velocity. A generalized form for the dynamics introduces a Lagrangian, L, which has the property that the equations of motion can be obtained from what are known as the Lagrange equations, which have a standard form:

$$\frac{d}{dt}\left(\frac{\partial L}{\partial \dot{z}}\right) - \frac{\partial L}{\partial z} = 0, \tag{1.10}$$

where z represents any position variable and $\dot{z}$ represents its time derivative (Killingbeck and Cole, 1971, 302). The Hamiltonian of a conservative system is then obtained from L by a standard procedure which involves what are known as canonical momenta, defined as the partial derivatives of L with respect to the $\dot{z}'$s and written as p'_zs. In most cases, L is the difference between the kinetic and potential energies, and the Hamiltonian comes out to be the total energy written in terms of position and canonical momentum variables. In the presence of an external field the Lagrangian needed to give the equations of motion when substituted into (1.10) takes a somewhat unusual form, and the canonical momenta are not equal to the usual expressions for momenta. The Hamiltonian, obtained by the standard procedure, then comes out to have a modified form for its kinetic energy part (see Van Vleck, 1944, chap. I). It is replaced by

$$\frac{1}{2m} \left(\mathbf{p} - e\frac{\mathbf{A}}{c} \right)^2 \tag{1.11}$$

where e is the charge on the electron and $\mathbf{A}$ is the magnetic vector potential. The latter is defined by a relation $\mathbf{B} = \text{curl } \mathbf{A}$, where $\mathbf{B}$ describes the field. $\mathbf{A}$ and $\mathbf{B}$ are concepts that are difficult to grasp, so a more detailed discussion of them will be given shortly. Here it is sufficient to accept that both are defined in terms of the field that would be present if the sample were to be removed without affecting the source of the field, and that there are different definitions for them, depending on the choice of units. (The inclusion of c indicates that the cgs form is being used.) The detailed analysis shows that $m\mathbf{v}$ is equal to $\mathbf{p} - (e\mathbf{A}/c)$, so the expression in (1.11) is actually equal to the usual kinetic energy. However, when the Hamiltonian is quantized it is $\mathbf{p}$ that is replaced by $(\hbar/i)$ grad, and not $m\mathbf{v}$.

In forming the quantum mechanical Hamiltonian it is also necessary to include a term for the energy of the spin in the magnetic field, which is sometimes denoted by $\mathbf{B}$ and sometimes by $\mathbf{H}$. One form for this interaction is $g\beta\mathbf{B} \cdot \mathbf{s}$, where β is the Bohr magneton, which in cgs units is $e\hbar/2mc$. (This use of β should not be confused with the β used for $1/kT$ in statistical mechanics.) In SI units β is replaced by μ_B, the Bohr magneton. The Dirac theory showed that g should be equal to 2, and whatever units are used, β and μ_B are chosen to reproduce this value, except that experiment has shown that the correct value is slightly different, being closer to 2.0023 (Lamb and Retherford, 1947). The reason for the extra 0.0023 is not easily explained, nor is it necessary to make the correction in most theoretical work. So the value $g = 2$ will be used throughout.

The physical origin of the departure from 2 is connected with the concept that the vacuum contains harmonic-like modes of oscillation of the electromagnetic field, each of which, even in its lowest-energy state, has zero-point energy. So a charged particle in free space is exposed to the electric fields associated with the zero point and any other excitations of these modes. Strictly speaking, the Hamiltonian should contain terms that describe the modes and their interactions with the electron. [There is also another effect, that any motion of the electron will tend to drag the electromagnetic (e.m.) field along with it, which will alter its perceived mass.] It is fortunate that it has been possible to show that for describing magnetic phenomena it is sufficient to use the experimental value of the mass and the observed 2.0023 value for g in what is really an effective Hamiltonian, one which omits all mention of the e.m. modes and their interactions with the electron.

The presence of **A** in the Hamiltonian has generated a good deal of interest. In dealing with Maxwell's equations in classical physics, a vector potential, which is usually denoted by **A**, is introduced. It is related to the vector **B** which describes the magnetic field by $\mathbf{B} = \text{curl}\mathbf{A}$. In this context **A** is regarded as an auxiliary variable which it has been convenient to introduce to help simplify the mathematics. It is not regarded as having physical significance. But when what seems to be the same quantity appears specifically in a quantum mechanical Hamiltonian, it becomes difficult to maintain that it has no physical significance. A comparison between its role in classical e.m. theory and in quantum mechanics shows that there are differences. In e.m. theory the wave motion is regarded as occurring in a microscopically continuous medium and then it is useful to introduce another vector, **H**. On the other hand, in quantum mechanics it is usual to assume that the motion of charged particles is taking place in a vacuum, and in using cgs units, the two fields that correspond to **B** and **H** are identical, so there is no need to have two apparently different vectors to describe the same thing. (In SI units, for some reason that escapes your author, the two vectors **B** and **H** are defined so that they have different dimensions and magnitudes.)

The question of how to relate the classical macroscopic concepts of **B** and **H** to the microscopic fields in a solid is discussed at some length in the opening pages of Van Vleck (1944). Van Vleck denotes the microscopic fields, fields in a vacuum, by **b** and **h** and concludes that **B** is the volume average of **h**. (From the analysis it is clear that there is an expectation that **h** will vary quite strongly within an atom, because it includes not only any external fields but also the fields from any internal microscopic magnets.) But what he does not do is explain how **h** can be measured and, from the point of view of quantum mechanics, this presents a problem, because it appears to assume that some microscopic probe is available that will not disturb the other magnetic moments in the solid.

From a variety of recent discussions it seems that there is still a good deal of confusion over **B** and **H** in macroscopic magnetism, so a course will be taken that seems to avoid the confusion. It will be assumed that there is no need to distinguish between **B** and **H**, for all the magnetic fields of interest will be fields in a vacuum. Either **H** or **B**, or even Van Vleck's **h** and **b**, can be used in a vacuum, for they all describe the same physical entity. The relation with **A** can be taken to be any one of **H**, **B**, **h**, or $\mathbf{b} = \text{curl}\mathbf{A}$. For convenience, though, **B** will be chosen, for this at least leaves H available to indicate the Hamiltonian. In adopting this stance, there is the important underlying assumption that **B** describes the magnetic field which would be there in a vacuum in the absence of the sample. All internal fields, such as those that in a classical picture are regarded as arising from internal magnetic moments, can be indirectly represented in the Hamiltonian in the form of interactions, energies between dipoles. These energies will thus be expressed through quantum mechanical operators, in contrast with the coupling of the system to an external magnetic field, where the field itself will be treated as a classical variable.

There is a further property of **A**, which makes it unusual. It is not uniquely defined in either classical or quantum physics. For the former this is of no significance, but for the latter **A** occurs explicitly in the Hamiltonian, so whether or not it has physical significance needs to be tested by experiment. This all comes about because the curl of the grad of any function is zero, so adding an arbitrary term, gradχ, to **A** makes no difference to its curl, though it does change the Hamiltonian. It would, of course, be disastrous if doing so changed any physical predictions. As far as eigenvalues are concerned, changing **A** by adding gradχ, where χ is independent of time, is equivalent to applying a unitary transformation, $\exp(ie\chi/\hbar c)$, to the Hamiltonian. This leaves the eigenvalues unaltered but changes the phases of the eigenfunctions. [The proof relies on the fact that $\exp(i\chi)(\mathbf{p})\exp(-i\chi) = (\mathbf{p} - \text{grad}\chi)$, where χ is any function of x, y, and z. It needs to be supplemented with a limitation on the allowed choices for χ, to ensure that the boundary conditions on the eigenfunctions after the transformation are consistent with those applied before it.] If **A** is time dependent, the analysis is more complicated, but as no examples where this may be relevant will be considered, this point will not be pursued.

With **A** appearing in the quantum mechanical Hamiltonian, there is still the question, over and above any ambiguity in its choice, of whether it is a measurable quantity. This was examined experimentally by using the possibility of having **B** zero and **A** nonzero. Thus Chambers (1960) showed that an electron traversing such a region experienced a change in the phase of its wave function, relative to that experienced when both **B** and **A** are zero, thereby establishing that **A** really does have physical significance. The experiment verified a prediction by Aharonov and Bohm (1959).

To avoid any possible ambiguity over the definition of **A** a specific choice will be made, that $\mathbf{A} = (1/2)(\mathbf{B} \wedge \mathbf{r})$. **A** is therefore invariant under any rotation of axes.

The "kinetic energy term" in the Hamiltonian can be expanded into an expression containing different powers of **B**, thus:

$$\left(\mathbf{p} + e\frac{\mathbf{A}}{c}\right)^2 = \mathbf{p}^2 + \frac{e\hbar}{c}\mathbf{B}\cdot\mathbf{l} + \frac{e^2}{4c^2}(\mathbf{B}\wedge\mathbf{r})\cdot(\mathbf{B}\wedge\mathbf{r}) \qquad (1.12)$$

where the charge on the electron has been written as $-e$, so that e is a positive quantity, and the above definition of **A** has been used. Adding the interaction of the electron spin with the magnetic field, the term linear in **B** becomes

$$\beta(\mathbf{l} + 2\mathbf{s}) \cdot \mathbf{B}.$$

There are a number of further terms which should be added, for the proton has a spin and a magnetic moment.

The structure of these terms will not be pursued here, for the reason for going this far is that the quantum mechanical treatment of hydrogen is unusual in that it can be taken much further, without approximations, than any other system. It can therefore be used to illustrate that while quantum mechanical Hamiltonians are ostensibly based on classical models, this is only partially true. The study of hydrogen has shown that there is sound evidence that nonclassical interactions exist. It is then reasoned that similar interactions will be present in more complicated systems, and, indeed, there is often experimental evidence which shows that extra interactions are required. It is, though, more a matter of faith than certainty that the forms chosen for them are correct, because it is only usually possible to verify them in the context of approximate treatments of the Hamiltonian.

§ 1.7 The Many-Electron Ion

In moving towards a treatment of the ionic magnetic crystals the next simplest systems to consider are the many-electron ions, for although it is not possible to go as far with them as with hydrogen, without approximations, the theory which has been developed is impressive in what it explains. It also provides a good basis from which to begin the study of ions in crystals.

It is invariably assumed that the nucleus of an ion is at rest, and that the nucleus contains a definite number, Z, of protons and a number of electrons that is not much different from Z. The Hamiltonian is clearly going to be more complicated than that of hydrogen because, apart from the increase in the number of electrons, it can be expected that there will be many smaller interactions. As well as the spin–own-orbit interactions, which have a counterpart in hydrogen, there are likely to be other magnetic interactions, such as spin-spin, spin–other-orbit, and orbit-orbit interactions. Even without these there is enough of a problem, for it is the presence of the Coulomb interactions between the electrons that has so far hindered progress in finding exact solutions. That they cannot be neglected, even as a first approximation, is quite clear, for there must be an approximate balance, for a specific electron, between the attraction it feels towards the nucleus and the repulsion it feels from all the other electrons. To make any progress a physically reasonable approximation is needed.

The favored one is to assume that the effect of all the electric forces, on a specific electron, is approximately that of a suitable static electric field with a smaller superimposed time-dependent field which is zero on a time average and so can be neglected in a first approximation. Then there is another point, that the exclusion principle requires the electrons to be indistinguishable, which can only be incorporated by assuming that they all experience the same field. It is therefore convenient to assume that the electric potential in which each electron moves depends only on r, its distance from the nucleus. The approximation has two attractive features. The first is that the approximate Hamiltonian is invariant under rotations of axes, and the second is that it separates into a set of hydrogenic-like Hamiltonians, one for each electron and all of the same form. So much of the theory can be taken over en bloc, for all that is changed is the radial dependence of the potential, which implies that the angular properties of the wave functions are the same as those found with hydrogen.

Various ways of estimating the potential, $V(r)$, have been investigated, but even without doing this quite a lot can be inferred. For example, it can be expected that the energy level pattern of each one-electron Hamiltonian will not be all that different from that of hydrogen, and that the main difference will be the disappearance of n as a quantum number which gives the energy and the degeneracy of each eigenvalue. There is now no reason to have the l degeneracies of a given n. Another use has therefore been found for n. In hydrogen the lowest state, $(1s)$, is nondegenerate. If $V(r)$ is not too different from that in hydrogen, it can be expected that the ground state will retain the s symmetry, so it can again be denoted by $(1s)$, where the $n = 1$ simply implies that it is the lowest energy level with s symmetry. There is rather more of a question over the next energy level, for in hydrogen the $(2s)$ and $(2p)$ coincide, and a similar degeneracy is hardly to be expected. However, for reasonable choices of $V(r)$ the energy levels which are next above $(1s)$ usually consist of a close pair, one a nondegenerate level with a state having s symmetry and the other a degenerate level of three states showing p symmetry. It is therefore reasonable to refer to the first of these as a $(2s)$ level, the second s level and to the other, with less reason, as a $(2p)$ level, for it is the first level with p symmetry. For the next set of states the labeling is again similar to that in hydrogen, with the first d-like state being labeled $(3d)$. The rule is that each level with a given l value is labeled in sequence by n, with the lowest n for a given l beginning at $(l + 1)$. [It would seem more logical to denote the first p level by $(1p)$, the first d level by $(1d)$, and so on. That this is not done dates back to the introduction of the present notation by the experimentalists, before the advent of quantum mechanics. In the theories of nuclear structure, where similar one-nucleon Hamiltonians are used, the first p level is denoted by $(1p)$ and the first d level by $(1d)$.] The $s, p, d, f, \ldots$ labeling, which also predates the theory, is an alternative to the $l = 0, 1, 2, 3 \ldots$ system. Both are still in use.

§ 1.8 The Exclusion Principle

At this stage three different Hamiltonians have been introduced to describe the many-electron ion. The first,

$$H_1 = \sum_i \left[\frac{\mathbf{p}_i^2}{2m} - \frac{Ze^2}{r_i} \right] + \frac{1}{2} \sum_{i \neq j} \frac{e^2}{r_{ij}}, \tag{1.13}$$

is an approximation to the generic Hamiltonian. It is made up of the interactions listed in (1.1), (1.2), and (1.3), with the restriction that only one nucleus is present. The second,

$$H_2 = \sum_i \left[\frac{\mathbf{p}_i^2}{2m} + V(r_i) \right], \tag{1.14}$$

is a many-electron Hamiltonian in which each electron moves in the same central potential, $V(r)$. The third,

$$H_3 = \left[\frac{\mathbf{p}^2}{2m} + V(r) \right], \tag{1.15}$$

is a one-electron Hamiltonian in which the electron moves in the same central potential as the electrons in H_2. The three Hamiltonians are related, and the next step is to examine this, building up the properties of H_2 from those of H_3 and some of those of H_1 from those of H_2.

At first sight, the lowest energy of H_2 might be expected to be equal to the lowest energy of H_3 multiplied by the number of electrons it describes in the summation over i. This, however, is not what it is found, and here it becomes necessary to introduce a concept that is now universally accepted as an essential feature of electronic systems, one which could not have been deduced using a classical model and one which is not apparent in the Hamiltonian formalism. It is a condition, imposed on all electronic wave functions, which was first proposed by Pauli, that no two electrons in a atom can have the same set of quantum numbers. Here it will be given in another form, that each many-electron wave function (where wave function is taken to include a description of the spin orientation) must reverse in sign under interchanges of the electronic labels (which also includes those of the spin) of any two electrons. (The interchange of labels means that the variables of electron i are to be replaced by those of electron j, and vice versa. "Antisymmetric" is used to describe the sign reversal. The use of "wave function" is not meant to imply that the state is an eigenstate of some operator, but simply that it has arisen in a mathematical analysis.) The antisymmetric requirement has the effect of eliminating many of the eigenstates of many-electron Hamiltonians, states which are of no physical significance. In the present context, H_2 has no allowed eigenstate with more than two electrons in $(1s)$, more than two in $(2s)$, and more than six in $(2p)$, etc.

To demonstrate the implications it is best to take an example. Suppose, therefore, that the ion of interest has seven electrons and that the eigenstates of H_3, when arranged in increasing order of energy, go as $(1s)$, $(2s)$, $(2p)$, ... Then the lowest-lying state of H_2 is obtained by having two electrons in $(1s)$, with opposite values for m_s, two in $(2s)$ with opposite m_s values, and three in $(2p)$. The energy eigenvalue of H_2 then becomes $2E(1s) + 2E(2s) + 3E(2p)$, in an obvious notation. There is, however, an ambiguity about the $(2p)$ arrangement, because there are actually six different states in $(2p)$ into which electrons can be placed; this arises because there are three m_l orbitals and two m_s possibilities. So the three electrons in $(2p)$ can be housed in $^6C_3$ different ways. Without the antisymmetric requirement it would seem that one of these arrangements could have the wave function

$$
\begin{aligned}
[\psi_{1s,0,+}(1)] \quad & \times \quad [\psi_{1s,0,-}(2)] \times [\psi_{2s,0,+}(3)] \times [\psi_{2s,0,-}(4)] \\
& \times \quad [\psi_{2p,1,+}(5)] \times [\psi_{2p,0,+}(6)] \\
& \times \quad [\psi_{2p,-1,+}(7)]
\end{aligned} \tag{1.16}
$$

where the first electron has been placed in the $(1s)$ state with $m_l = 0$ (the only possible value, with l added as a subscript to distinguish it from m_s) and $m_s = 1/2$ (denoted by $+$). The second has been placed in the same orbital but with $m_s = -1/2$ (denoted by $-$), and so on. For electrons 5, 6, and 7, specific choices have been made for the m_l and m_s values from the three possibilities of 1, 0, and -1 for m_l and the two possibilities, $+$ and $-$, for m_s. An interchange of electrons 1 and 2 will produce an orthogonal state of the same energy, with the orthogonality arising because in (1.16) the first electron is in a state with $m_s = 1/2$ and after the interchange it has $m_s = -1/2$. This interchange clearly does not give a state that is the state in (1.16) with a reversed sign.

There is an elegant way of obtaining a state which does have the required antisymmetry from a state such as that given in (1.16). It relies on a property of determinants. The first step is to set up a determinant in which the states $\psi_{1s,0,+}(1)$, $\psi_{1s,0,-}(2)$, etc., are placed in sequence down its principal diagonal, and to follow it with a systematic procedure for determining the rest of the elements. Beginning with the top row, the element on the far left contains (1), the label for the coordinates of the first electron. In the element to its immediate right, (1) is replaced by (2), so it is an expression in the coordinates of the second electron. In the position to the right of (2) the 2 is replaced by 3, and so on across the row. In the second row it is the second element which contains the 2. So in the first element of this row 2 is replaced by 1, and all the other elements in the row are generated by running though the sequence 2, 3, ... in place of the 1 for the first element, as for the first row. Continuing in this way, all the elements can be obtained from the specification of the diagonal elements. (There is an alternative, to run down columns instead of rows.) If the determinant is now expanded it is found that it consists of 7! orthonormal states, each of which has a structure similar to that given in (1.13). A well-known property of determinants, that interchanging any two rows or columns reverses its sign, ensures that the state is antisymmetric. Another property, that determinants vanish if two or more rows or columns are identical, shows that there are no antisymmetric states in which two or more electrons have the same quantum numbers.

It has been assumed implicitly that each one-electron state is normalized, so the determinantal state needs to be divided by $(7!)^{1/2}$ to be normalized. It is standard practice to include such a normalizing factor in the construction of the determinant, after which it is known as a Slater determinant. If the states in (1.16) had been written in some other order, or the electrons had been numbered in a different way, the resulting Slater determinant might have had a sign which is the opposite of that found beginning with (1.16). This would have been of no significance, for the phase of a wave function carries no physical information. It is, perhaps, worth stressing that although each of the 7! product states in the expansion of the Slater determinant is an eigenstate of H_2, only the determinantal combination is a state of physical importance. (It is sometimes useful to know that the Slater procedure can also be used to construct antisymmetric many-electron states from nonorthogonal one-electron states.)

The sequence of diagonal elements produces a determinantal function which is unique to within a phase factor. Placing seven electrons, two in $(1s)$, two in $(2s)$, and three in $(2p)$, allows the construction of twenty orthogonal determinantal states, according to how the electrons occupy the $(2p)$ states. All twenty will be eigenstates of H_2 and will have the same energy. This energy level, and indeed most of the other energy levels formed by rehousing the electrons in other choices of one-electron states, will be degenerate. The energies of any of these states can be found from the energy levels of H_3 by adding up the energies of the individual electrons.

The set of $^6C_3$ antisymmetric states is an example of what is known as a "configuration," a specification of the number of electrons in each hydrogenic orbital. For the present example it would be written as $(1s)^2(2s)^2(2p)^3$, where the superscripts give the number of electrons in each family of hydrogenic orbitals. For any ion, each state of a configuration is an eigenstate of an H_2, each one is antisymmetric, and together they form an orthonormal set of many-electron states all with the same eigenvalue, a sum of eigenvalues of H_3.

§ 1.9 The Approximate Eigenstates of H₁

H_2 and H_3 were introduced as stepping stones towards an understanding of the properties of H_1. Other than this they are of no interest, for they do not describe any physical system. It may not, however, be immediately obvious that there is anything important about H_1 to be learnt from them. The answer is that there is, provided one has access to the right mathematical tools. What these should be is not always initially apparent, though the position does tend to become clearer, strangely enough, when attempts are made and the wrong ones are eliminated. In the next paragraph an example of what can be done with limited resources will be given. It indicates a direction for further mathematical developments, which will be even further developed in some of the following chapters.

Still using the seven-electron example and remaining with its lowest-energy configuration, it is clear that there is no scope for rearranging electrons in the $(1s)$ and $(2s)$ shells. However, the $(2p)$ orbitals are only partially filled. It is then of interest to consider whether the H_2 result that all the possible rearrangements have the same energy looks reasonable on physical grounds. For example, it has eigenstates in which two of the electrons are in the $(2p_x)$ orbital state, with opposite spins, with the third electron in either $(2p_y)$ or $(2p_z)$. One can ask whether the charge distribution will be the same or different if one of the electrons in $(2p_x)$ is moved into $(2p_y)$. The answer depends on whether or not the third electron is, or is not, already in $(2p_y)$. If it is, then the electron to be moved will need to have the opposite spin. The result will be a charge distribution that is a rotated version of the initial arrangement. On the other hand, if $(2p_y)$ is initially unoccupied then the third electron must be in $(2p_z)$ and the final arrangement will have one electron in each of p_x, p_y, and p_z. It is by no means obvious that the two possible final charge distributions have the same electrostatic energy, in spite of H_2 indicating that they have the same eigenvalues. The disparity has arisen because in H_2 the Coulomb interactions between the electrons have been replaced by an effective potential common to all the electrons, whereas a nonmathematical picture has suggested that a feature of physical significance has, probably, been overlooked.

Since both H_1 and H_2 describe systems with the same number of electrons it is possible to define a fourth Hamiltonian:

$$H = H_2 + \lambda(H_1 - H_2),\qquad(1.17)$$

where λ is a parameter that can be varied from 0 to 1. In the process H will change from H_2 to H_1, so if the properties of H for a general value of λ could be established, it would be possible to see how the properties of H_1 are related to those of H_2. The theory needed to do this is known, generally, as perturbation theory. It is not, however, just one mathematical technique, and its description, even in the limited field of magnetic ions, needs a separate chapter. In the present case, it so happens that further progress can be made without needing perturbation theory.

H_1 and H_2 have spherical symmetry, invariance under rotation of axes, and therefore so does the H of eqn. (1.17). On varying λ from 0 to 1 and using classical concepts, the total angular momentum of the system will be conserved, because the forces which are being changed as λ increases are either equal and opposite internal forces or forces that go through the origin. With $\lambda = 0$ the electrons are uncoupled, so their total angular momentum is the vector sum of the angular momenta of the individual electrons. This will be independent of the value of λ. So with a nonzero λ the motion is likely to be one in which no electron has a constant value of angular momentum; there will be interchanges of angular momentum between the electrons in such a way that their total angular momentum remains constant.

When the same problem is examined in quantum mechanics, it is found that the total orbital angular momentum, **L**, which is the sum of the individual **l**'s, commutes with H for all values of λ. It is therefore a constant of the motion. However, unlike a classical example there is an additional source of angular momentum, in the electron spins. The total spin angular momentum, **S**, the sum of the individual **s**'s, also commutes with H and so it too is conserved as λ changes.

It will be convenient to continue by treating **L** and **S** as independent operators. However, before doing so it will be useful to have available some general properties of angular momentum operators.

The angular momentum operators for different electrons commute, so it readily follows that **L** and its components obey the same commutation rules as the individual **l** operators. Thus $\mathbf{L} \wedge \mathbf{L} = i\mathbf{L}$ and $\mathbf{L} \cdot \mathbf{L}$ commutes with each component of **L**. Also, all components of **L** commute with H, which makes it convenient to classify its eigenstates as eigenstates of $\mathbf{L} \cdot \mathbf{L}$ and L_z. The possible eigenvalues of $\mathbf{L} \cdot \mathbf{L}$ can be written as $L(L+1)$, where L is a positive integer or zero and the eigenvalues of L_z take all positive and negative integer values between $-L$ and L, including 0. As with the lower-case operators, it is useful to introduce ladder operators

$$L_+ = L_x + iL_y \ \text{ and } \ L_- = L_x - iL_y,$$

for on denoting a simultaneous eigenstate of $\mathbf{L} \cdot \mathbf{L}$ and L_z, which have the respective eigenvalues $L(L+1)$ and M, by $|L, M\rangle$ it can be proved that

$$L_+|L, M\rangle = [L(L+1) - M(M+1)]^{1/2}|L, M+1\rangle$$

and

$$L_-|L, M\rangle = [L(L+1) - M(M-1)]^{1/2}|L, M-1\rangle.$$

These equations are analogous to those for the l_+ and l_- operators acting on one-electron states (see §1.4). (Similar relations hold for the spin operators, where S and M_S are usually used in denoting the spin eigenstates, in which case M for the orbital moment is usually also given a subscript and written as M_L. There is a slight difference, however, for S and M_S can take either integer or integer plus one-half values according to whether the number of electrons is even or odd (see Edmonds, 1968; Brink and Satchler, 1968; and Thompson, 1994).

For the ground level of the seven-electron system it is first supposed that $\lambda = 0$ and that the electrons are arranged to have the maximum possible value of L_z. The electrons in filled shells can be ignored, for there is no flexibility in their arrangement and they have zero total orbital and total spin angular momentum. Only the outer three electrons need to be considered, and by inspection the largest M_L is obtained by letting the first electron have $m_l = 1$ with m_s either $1/2$ or $-1/2$, the second have $m_l = 1$ and the opposite value for m_s, and the third have $m_l = 0$ with m_s either $1/2$ or $-1/2$. Such a state has $M_L = 2$. It can then be deduced that since this is an arrangement which has the largest possible L_z value, the rules that govern angular momentum show that it must be an eigenstates of $\mathbf{L} \cdot \mathbf{L}$ with L equal to 2. (If an L greater than 2 occurred it would have a state with M_L greater than 2, and there is no such state.) The state can be written as an eigenstate of L and M_L:

$$\{1, 0, 0, \quad + \quad : 1, 0, 0, - : 2, 0, 0, + : 2, 0, 0- : 2, 1, 1, + : 2, 1, 1,$$
$$- \quad : 2, 1, 0, +\} = |2, 2, \rangle, \tag{1.18}$$

where the use of $\{\ldots\}$ on the left is meant to indicate that the state is a Slater determinant. (It is only necessary to list the elements on its diagonal. Each element has values for four symbols, n, l, m_l, and m_s, and those of one element are separated from those of another by a colon.) The state on the right is the same state in L, L_z quantization.

The notation is incomplete because definite m_s values have been chosen for the states on the left-hand side and there is no corresponding spin description in the state on the right. The omission will be corrected shortly. If L_+ is applied to the right-hand side it produces zero and so, therefore, should $\sum l_+^i$, a summation over all electrons of the one-electron orbital raising operator, when applied to the left-hand side. What it actually does is to give a sum of determinants, as if l_+ acted successively on each one-electron state. The first four of these all have $l = m = 0$, so the result is four zeros, because l annihilates each state in turn (l_+ applied to a state for which $l_z = m$ introduces a factor $[l(l + 1) - m(m + 1)]^{1/2}$ that vanishes when $l = m = 0$). The next two also vanish, for l_+ on a state with $l = 1$ and $m_l = 1$ also gives zero. The seventh, which has $l = 1$, $m = 0$, produces

$$\sqrt{2}\{1, 0, 0, \quad + \quad : 1, 0, 0, - : 2, 0, 0, + : 2, 0, 0, - : 2, 1, 1,$$
$$+ \quad : 2, 1, 1, - : 2, 1, 1, +\},$$

which vanishes for a different reason, that two electrons are in the same state. So $\sum l_+^i$ applied to the left-hand side does indeed give zero.

It is also possible to apply S_+ to the left-hand side. Using the same procedure, each of the seven determinants vanishes because either the starting one-electron state has $m_s = 1/2$ or it has $m_s = -1/2$, which, when changed to $1/2$, results in two electrons being in the same state. This demonstrates that the $L = 2$, $M_L = 2$ state, which clearly has $M_S = 1/2$, must also have $S = 1/2$. If S_- is applied instead, the first six determinants again vanish, but the seventh gives

$$\{1, 0, 0, \quad + \; : 1, 0, 0, - : 2, 0, 0, + : 2, 0, 0, - : 2, 1, 1,$$
$$+ \; : 2, 1, 1, - : 2, 1, 0, -\},$$

which is therefore a state with $L = 2$, $M_L = 2$, $S = 1/2$, and $M_S = -1/2$. It can now be seen how the description of the right-hand states in (1.18) can be improved, by writing it as $|2, 2, 1/2, 1/2\rangle$, a sequence of L, M_L, S, M_S quantum numbers. Similarly, the normalized state obtained by applying S_- to it can be written as $|2, 2, 1/2, -1/2\rangle$. It can also be seen that the notation for the Slater determinants would have been simplified, without loss of information, if the details about the electrons in filled shells had been omitted. Such omissions are common.

Having established that the seven-electron state on the left-hand side of eqn. (1.18) has $L = 2$ and $S = 1/2$, it can now be inferred that there must be a family of ten many-electron states all of which have these values for L and S but which differ in their M_L and M_S values, the possibilities for M_L being $2, 1, 0, -1$, and -2 and for M_S, being $1/2$ and $-1/2$. Such a family is called a "term," and it is conventionally denoted by ^2D, where the superscript equals $(2S + 1)$ and D denotes $L = 2$ (an extension of the notation for a d electron, which has $l = 2$). It has therefore been possible to deduce, even when λ is nonzero and exact eigenstates cannot be obtained, that there will be a degenerate energy level, labeled ^2D, which is spanned by ten orthonormal many-electron eigenstates. When λ is zero all the states in this term can be determined by applying the L_+, L_-, S_+, and S_- operators to either of the above determinantal states, though this is not recommended, except as an exercise, for most of them are rather lengthy expressions which are seldom needed. When λ is nonzero the analysis gives only a limited amount of information about the eigenstates, but enough to show that there will be at least ten states with the same energy.

The analysis can be continued by looking for the state that has the maximum value for M_S, ignoring the electrons in filled shells. It occurs when all three $(2p)$ electrons have $m_s = 1/2$, which restricts the orbital choice to the m_l's of 1, 0, and -1. So there is only one such state, and it must have $S = 3/2$, $M_S = 3/2$, $L = 0$, and $M_L = 0$. There is no scope to have related states by allowing M_L to take other values, but there is scope for varying M_s, to $1/2$, $-1/2$, and $-3/2$. There will therefore be a family of four states with the same energy, for any value of λ. This is described as a ^4S term.

With three electrons in a p shell, there are twenty orthonormal Slater determinants, and so far it has been deduced that with $\lambda = 0$ it is possible to take orthonormal combinations to produce ten states in a ^2D term and four states in a ^4S term. The final step is to look at the combinations that are left. The only possibilities for M_L are 1, 0, and -1 and for M_S 1/2 and $-1/2$. These are the values which go with a ^2P term. It has therefore been established that, when λ is nonzero, there will be twenty states arising from the ground configuration of the seven-electron system, which can be regarded as falling into three terms, with the states in each term having the same energy. The analysis has said little about the actual eigenstates and nothing about their energies, though there is a strong implication that each term will have a different energy. The best approximations to the eigenvalues of H_1 to which this argument leads will be obtained by setting λ to unity and evaluating the expectation value of H_1 over some convenient state in each term found with $\lambda = 0$. If such a state happened to be an eigenstate of H_1, which is actually unlikely, the expectation value so obtained would be equal to the eigenvalue. However, before going on to give an example of what determining the expectation value involves, it may be useful to know that in Condon and Shortley's book there are tables of the terms for all the configurations that are likely to be of interest.

The calculation of expectation values of H_1 is simplified by means of rules for obtaining the matrix elements of operators between Slater determinants. The first step is to separate the operators according to whether they are sums over one-electron operators or sums over two-electron operators (interactions involving more than two electrons seldom occur in Hamiltonians). An example of the first type is the kinetic energy of the electrons, which is a sum over i of $\mathbf{p}^2/2m_i$, and an example of the second type is their mutual potential energy, which is usually taken as a sum over $i \neq j$ of $e^2/2\mathbf{r}_{ij}$. If U denotes a one-electron operator and $\{a, b\}$ and $\{c, d\}$ represent Slater determinants for a two-electron system, where $a, b, c,$ and d are orthonormal one-electron states, orbital plus spin, the requirement is to simplify the matrix element of $U(1) + U(2)$ taken between the two determinantal states. In doing so it is convenient to replace a state such as a by its bra or ket form $|a\rangle$ or $\langle a|$. By expanding the determinant the element is found to equal

$$\left(\tfrac{1}{2}\right) [\langle a|\langle b| - \langle b|\langle a|][U(1) + U(2)][|c\rangle|d\rangle - |d\rangle|c\rangle], \qquad (1.19)$$

where in each $\langle \ldots |\langle \ldots |$ the first ket refers to electron 1 and the second to electron 2, and similarly in $|\ldots\rangle|\ldots\rangle$. (This is asymmetric, for it might be expected that the ket associated with a bra in which the electrons are ordered from 1 to 2 would have the electrons ordered from 2 to 1. The conventional arrangement has the usual sequential ordering.) It is then seen that the matrix element vanishes if a, b, c, and d are all different. But if $a = c$ and $b \neq d$ the element reduces to $\langle b|U|d\rangle$ and if $a = c$ and $b = d$ it is equal to $\langle a|U|b\rangle + \langle c|U|d\rangle$. The reasoning, which is readily extended to a system with more than two electrons, yields the following rules. The matrix element is zero if on comparing the two determinants they differ in two or more one-electron states. If they differ in only one, the order of the states in the determinants should be changed so that identical states occur in identical positions, a process which may introduce a factor of -1. Then the required matrix element is simply that of U between the two one-electron states which differ. If the determinants do not differ in their states, then they should be reordered to have identical states in identical positions, thus possibly introducing a factor of -1. The matrix element is then the sum $\langle i|U|i \rangle$ over all i. (The actual many-electron states are not simple determinants, or sums of determinants, because each Slater determinant has a numerical normalizing factor. These are ignored in using the rule. It is as if the rule is applied directly to the $\{\ldots\}$ form. In evaluating (1.19) the normalizing factors of $(1\sqrt{2})^2$ have been included. They are canceled by a factor of 2 because, in simplifying the expression, each nonzero matrix element occurs twice.)

For the two-electron operators it is again best to begin with a two-electron example and simply substitute the two-electron operator, V_{12}, for the two one-electron operators in eqn. (1.19). The matrix element is found to equal

$$\langle a, b|V_{12}|c, d \rangle - \langle a, b|V_{12}|d, c \rangle \tag{1.20}$$

when use is made of a symmetry property of V, that it is invariant under interchanges of the labeling of the two electrons. With more than two electrons, say N, it is readily apparent that there will be no matrix elements between determinantal states unless they have at least $N - 2$ one-electron states in common. These states should first be ordered to put identical states in corresponding positions in the determinants. There are then three possibilities. The first is that the two determinants have only $N - 2$ states in common. The states should then be ordered, with sign changes if necessary, so that all those in common in the two determinants are either to the far left or to the far right and similarly ordered. The matrix element is then identical with that for the two-electron example, when the parts in common are ignored. That is,

$$\langle a, b, \ldots, c, d|\tfrac{1}{2}\sum_{i \neq j} V_{i,j}|a, b, \ldots, e, f \rangle = $$
$$\langle c, d|V_{1,2}|e, f \rangle - \langle c, d|V_{1,2}|f, e \rangle. \tag{1.21}$$

If the two determinants differ in only one state, then this should be moved to either the far right or the far left, and the rest ordered so that identical states are in corresponding positions. Comparing this matrix element with the one on the left in eqn. (1.21), it is apparent that it differs in that c is now identical with e, so the matrix element could be identified with that in (1.20). However, if the states in the two bras and kets that have elements in corresponding positions were both reordered, any one of their states could appear in the position occupied by c. So there will be matrix elements in which c is replaced by any of the states that are common to the two determinantal states. The overall value of the matrix element is thus equal to

$$\sum_c [\langle c, d | V_{1,2} | c, f | - \langle c, d | V_{1,2} | f, c \rangle], \tag{1.22}$$

there being $N - 1$ choices for c. When both determinants have all N states in common they should again be similarly ordered. The matrix element is then equal to

$$\tfrac{1}{2} \sum_{c \neq d} [\langle c, d | V_{1,2} | c, d \rangle - \langle c, d | V_{1,2} | d, c \rangle]. \tag{1.23}$$

The rules for the two-electron operators can be summed up as follows. The matrix element between two determinantal wave functions for an N-electron system can be regarded as composed of expressions involving matrix elements such as would be found for a two-electron system, the nature of the sum being determined by how many of the electrons in the two determinants are in different states. If the determinants differ by two states, then there is only one expression in the summation. If they differ by one state, there are $N - 1$ expressions, and if they do not differ at all then there are $N(N-1)$ expressions. [The two electrons should be given labels to distinguish between them. 1 and 2 have been chosen. As V_{12} is invariant when 1 and 2 are interchanged, this symmetry can be used to reduce the number of expressions to $(1/2)N(N-1)$.] Each expression has the form

$$\langle cd | V_{12} | ef \rangle - \langle cd | V_{12} | fe \rangle.$$

For the seven-electron example it is not possible to complete the evaluation of the matrix elements when V_{12} is put equal to e^2/r_{12}, because the radial parts of the one-electron states are unknown. It is nevertheless interesting to find that the signs of all the required two-electron matrix elements can be established, and that this makes it possible to show that ^4S is the term of lowest energy, ^2D comes next, and ^2P is the highest. (There is an approximation in this, for the so-called energies are expectation values of H_1 taken over the eigenstates of H_2.)

That the signs can be determined comes about because the angular parts of the states in the various one-electron states are known and only two different types of two-electron matrix elements are needed. The first of these is known as a Coulomb matrix element. It has the form

$$\langle a, b | \frac{e^2}{r_{12}} | a, b \rangle \quad \text{or} \quad \int \psi_a^*(1) \psi_a(1) \frac{e^2}{r_{12}} \psi_b^*(2) \psi_b(2) d\tau_1 d\tau_2, \tag{1.24}$$

where, to avoid the matrix element on the left vanishing, it has been assumed that the state in the bra labeled $a(b)$ has the same spin as the state labeled $a(b)$ in the ket. The integral form is recognizable as the energy of two positive charge distributions, each of which contains precisely one electron (the one-electron wave functions are normalized). So the matrix element is positive. The other type of matrix element has the form

$$\langle a, b | \frac{e^2}{r_{12}} | b, a \rangle \quad \text{or} \quad \int \psi_a^*(1)\psi_b(1)\frac{e^2}{r_{12}}\psi_b^*(2)\psi_a(2)d\tau_1 d\tau_2, \qquad (1.25)$$

where it is assumed that the first state a in the bra vector has the same spin as the first state b in the ket vector, and similarly for the second states, otherwise the integral vanishes. It represents the mutual Coulomb energy of a charge distribution given by the product $\psi_a^*\psi_b$ with another charge distribution given by the charge distribution $\psi_a\psi_b^*$. Unlike the Coulomb matrix element, each charge distribution has a total charge of zero, because the orthogonality of ψ_a and ψ_b requires the integral of $\psi_a\psi_b^*$ over all space to vanish. Such a matrix element, which is known as an exchange matrix element, is also positive. The proof uses the expansion

$$\frac{1}{r_{12}} = \frac{1}{|\mathbf{r}_1 - \mathbf{r}_2|} = \frac{4\pi}{r_>(2l+1)} \sum_{l,m} \left(\frac{r_<}{r_>}\right)^l Y_l^m(1)[Y_l^m(2)]^*, \qquad (1.26)$$

where $r_<$ is the lesser and $r_>$ the greater of $|\mathbf{r}_1|$ and $|\mathbf{r}_2|$ and the $*$ on the second Y denotes the complex conjugate. The positive sign follows on integrating term by term.

The conclusion that the term with the largest S has the lowest energy is an example of one of Hund's rules (Hund, 1925), which states that in any configuration the term of maximum S has the lowest energy. The example chosen is a fairly simple one; in more complicated cases, particularly with partially filled d or f shells (open shells), the same largest S may occur in several terms. The rule then states that the lowest term has, first, the maximum S and then the maximum L. This rule was deduced by examining many sets of experimental observations. It is interesting that the present comparatively simple theoretical treatment, which certainly involves approximations, does so well.

The apparent agreement of this example with Hund's rule demonstrates that there may be an important consequence of the exclusion principle. It seems that low energies are to be associated with electrons with parallel spins, electrons that are in different orbital states, for it is only then that exchange matrix elements, positive quantities, enter, and then they do so with negative coefficients.

§ 1.10 The Variational Principle

The determination of the radial parts of the one-electron hydrogenic wave functions, even approximately, is outside the scope of the theory that has so far been presented. Indeed, it could be regarded as a distraction, because the dominating interest is in dealing with the generic Hamiltonian, not approximations of the H_3 and H_2 type. Nor is it obvious that the concept has any physical reality. Nevertheless, it has been used a good deal and it does seem to have helped in understanding ionic spectra and a variety of other phenomena. Among the various procedures is one that is equally applicable in principle to the generic Hamiltonian, so it will be described in this context. It relies on a variational principle.

It was shown (see Slater, 1968) that if a guess is made of the normalized eigenstate of the lowest energy level of any Hamiltonian, the expectation value of the Hamiltonian, taken over that guessed state, will not be less than the true eigenvalue. Thus to obtain a best estimate for the ground state energy of H_1 it is possible to envisage using one of the states in the ground term of H_2 in an examination of how the expectation value of H_1 changes as this state is altered. So the technique is to make what seems to be a reasonable choice for $V(r)$, letting it contain a few parameters, and find the eigenstates and energies of the one-electron H_3. This information is then used to construct a ground state of H_2, for the radial parts of the one-electron wave functions are then known. The expectation value of H_1 can then be determined. The whole procedure can be repeated with different values for the parameters in $V(r)$. The minimum expectation value ought then to be a good approximation to the ground state energy of H_1. The method can be extended to determine the minimum energies of some of the higher-lying energy levels if they have states that remain orthogonal to the ground state as the form of $V(r)$ is changed. Energy differences, quantities which are more important than absolute energies, can then be estimated and compared with experimental values. It is difficult to decide whether, when this variational procedure gives a good approximation to an eigenvalue of H_1, it also gives a good approximation to the eigenstate.

There are a number of books, Judd (1963) and Woodgate (1970), for example, besides the one by Condon and Shortley, that describe the theory of many-electron ions.

§ 1.11 The Agreement between Theory and Experiment

In spite of the elegance of the above theoretical methods the theory of many-electron ions has to be taken a lot further before it is possible to make a proper comparison between experiment and theory. This is because spectroscopic observations on free ions have very high resolution, so separations between energy levels that are far smaller than the separations which the above theory would suggest are readily observed. Indeed, the spectra really show no direct evidence for the existence of terms. The reason is not hard to find. The terms are degenerate, the H_1 Hamiltonian has omitted a variety of small interactions which should be there, and the resolution of degeneracies, to give small splittings, is the most striking characteristic of small terms in a Hamiltonian. The theory has therefore been extended, to take account of a variety of magnetic interactions between the electrons and, where appropriate, the interactions between a nuclear spin and the electrons, which may contain both a magnetic and an electrostatic part, if the nucleus is nonspherical.

In tables of ionic energy levels it is usual to omit the splittings due to nuclear interactions. This does not mean that they are unobservable or that it is possible to switch them off. The implication is rather that with the aid of the theory it is possible to work out what the energy level pattern would have been if the nucleus had happened to have a spin of zero. This illustrates an important difference between theory and experiment, that the theoretician can switch off interactions by omitting them from the Hamiltonian, whereas the experimentalist has to put up with what is there. Nevertheless, it is possible, with a good enough theory, to work back from an observed energy level pattern to that which would have been observed if some of the interactions had been switched off. It is in this sense that theory has been able to account for term structures.

The theory that gives the terms has actually made an assumption about the magnitudes of the interactions in the generic Hamiltonian, for it has left out the magnetic interactions in favor of the electrostatic ones. This is only appropriate for certain ions. A general consideration of the various possibilities has led to the classification of ions as belonging to one of three classes: Russell-Saunders coupling, j-j coupling, and intermediate coupling. For the magnetic ions of most interest, those which have partially filled $3d$ and $4f$ shells, Russell-Saunders coupling is appropriate. For these ions, the electrostatic interactions are certainly larger than the magnetic ones. However, the latter are not negligible and they lead to a coupling between the total orbital and total spin momenta. This is usually dealt with by including an interaction, in H_2, of the form $\lambda \mathbf{L} \cdot \mathbf{S}$, where λ is known as the spin-orbit coupling parameter. The $\mathbf{L}$ of the orbital motion is then coupled to the $\mathbf{S}$ of the spins to produce a "total" angular momentum vector $\mathbf{J}$ equal to $\mathbf{L} + \mathbf{S}$. The components of $\mathbf{J}$ have the usual angular momentum commutation rules so that $\mathbf{J} \cdot \mathbf{J}$ comes out to have the form $J(J+1)$, where J ranges from $L + S$ to $|L - S|$. Using

$$2\mathbf{L} \cdot \mathbf{S} = (\mathbf{L}+\mathbf{S}) \cdot (\mathbf{L}+\mathbf{S}) - \mathbf{L} \cdot \mathbf{L} - \mathbf{S} \cdot \mathbf{S} = J(J+1) - L(L+1) - S(S+1)$$

it follows that the degeneracies associated with a term will be split into a pattern of levels, each of which is described as a multiplet, with each multiplet being characterized by a particular value of J and having $(2J+1)$-fold degeneracy. A similar conclusion is arrived at by including spin-dependent interactions, which then take a more complicated form, in H_1.

When the nucleus has a spin, $\mathbf{I}$, another angular momentum operator, $\mathbf{F} = \mathbf{J}+\mathbf{I}$, is introduced. The $(2J+1)(2I+1)$ degeneracy when the electron-nuclear

34 – CHAPTER 1

interactions are omitted gives a set of separated F-levels when it is included. As might be expected, being typical of angular momentum couplings, this latest angular momentum operator has an $\mathbf{F} \cdot \mathbf{F} = F(F + 1)$, where F ranges from $(J + I)$ in integer steps to $|J - I|$.

A valuable technique, for all ions, is to apply a magnetic field, for this generally results in all remaining degeneracies being lifted. Accounting for the arrangement of the energy levels and how they alter with the magnitude of the field is a sensitive test of the theory.

§ 1.12 References

Aharonov, Y., and Bohm, D. 1959. *Phys. Rev.* 115:485.

Brink, D.M., and Satchler, G.R. 1968. *Angular Momentum* (University Press: Oxford).

Chambers, R.G. 1960. *Phys. Rev. Lett.* 5:3.

Condon, E.U., and Shortley, G.H. 1967. *The Theory of Atomic Spectra* (University Press: Cambridge).

Dirac, P.A.M. 1947. *The Principles of Quantum Mechanics* (Clarendon Press: Oxford).

Edmonds, A.R. 1968. *Angular Momentum in Quantum Mechanics* (University Press: Princeton).

Hund, F. 1925. *Zeits. Phys.*, 33:855.

Judd, B.R. 1963. *Operator Techniques in Atomic Spectroscopy* (McGraw-Hill: New York).

Killingbeck, J., and Cole, G.H.A. 1971. *Mathematical Techniques and Physical Applications* (Academic Press: New York).

Lamb, W.E., and Retherford, R.C. 1947. *Phys. Rev.* 72:241.

Pauli, W. 1927. *Zeits. Phys.*, 43:601.

Slater, J.C. 1968. *Quantum Theory of Matter* (McGraw-Hill: New York).

Thompson, W.J. 1994. *Angular Momentum* (John Wiley and Sons: New York), §7.1.4.

Uhlenbeck, G.E., and Goudsmit, S. 1925. *Naturwiss.* 13:953.

Van Vleck, J.H. 1926. *Quantum Principles and Line Spectra* (National Research Council: Washington).

Van Vleck, J.H. 1932 and 1944. *The Theory of Electric and Magnetic Susceptibilities* (University Press: Oxford).

Van Vleck, J.H. 1935. *J. Chem. Phys.* 3:807.

Woodgate, G.K. 1970. *Elementary Atomic Structure* (McGraw-Hill: London).

References Providing Entry into the Literature on ESR

Some regularly produced journals are:

The Bulletin of Magnetic Resonance
Electron Spin Resonance, Specialist Periodical Reports, The Royal Society of Chemistry

The Journal of Magnetic Resonance
Magnetic Resonance Reviews

A recent book is: *Transition Ion Paramagnetic Resonance*, by J.R. Pilbrow (1995) (Clarendon Press: Oxford).

2

Group Theory

§ 2.1 Representations

Group theory is a branch of pure mathematics dealing with concepts that often seem too abstract for a typical physicist. A common reaction is to feel that it is difficult to comprehend because it is unclear where it is going, an impression which is not helped by the extensive literature on the subject, most of which is not used in theoretical physics. Nevertheless, some of its ideas are extremely useful. This account, which does not aim to be a detailed account of group theory, should therefore be regarded as an attempt to introduce a selection of concepts in a way that keeps close to their use in solid state physics. Since crystal field theory provides a good many examples of how these are used, it is convenient to use examples chosen with this in mind.

It is convenient to begin by considering a perfect cube, with side $2a$. It has eight vertices which can be regarded as at $(\pm a, \pm a, \pm a)$, using its center as origin and cartesian axes that pass through the centers of its faces. (See fig. 2.1.) The eight vertices can be numbered from 1 to 8, a convenient choice being to have opposite corners summing to 9. If now the cube is rotated so that the various vertices interchange, then in one sense the cube has been left unaltered and in another there has been a change because the numbering of the vertices has been altered. Such a rearrangement can be represented by an 8×8 matrix, the elements of which are either 0 or 1. To save having to write down an 8×8 matrix to illustrate this, a 3×3 example will be used, which can be pictured as arising from an equilateral triangle in a plane, which is rotated clockwise through $2\pi/3$. (See fig. 2.2.) The change in the sequence as the 1, 2, 3 numbering is replaced by 2, 3, 1 can be represented either as $1 \rightarrow 2, 2 \rightarrow 3$, $3 \rightarrow 1$ or in a matrix form:

$$\begin{pmatrix} V_1 \\ V_2 \\ V_3 \end{pmatrix} \rightarrow \begin{pmatrix} V_2 \\ V_3 \\ V_1 \end{pmatrix} = \begin{pmatrix} 0 & 1 & 0 \\ 0 & 0 & 1 \\ 1 & 0 & 0 \end{pmatrix} \begin{pmatrix} V_1 \\ V_2 \\ V_3 \end{pmatrix}, \qquad (2.1)$$

where V_i represents the vertex initially at site i. The 3×3 matrix can be regarded as representing the rotation, R, for if some different initial numbering or notation had been used the same 3×3 matrix would have resulted. R, regarded as a rotation, has the property that after three such rotations the initial

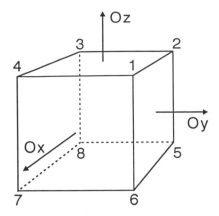

Figure 2.1. A possible labeling for the vertices of a cube.

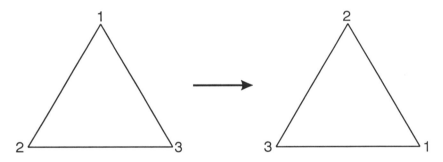

Figure 2.2. The vertices of an equilateral triangle are relabeled by a clockwise rotation through 2/3.

arrangement is restored, a result which can be written as $R^3 = E$, where R is the 3×3 matrix in (2.1) and E is a 3×3 unit matrix. Another operation is the reflection, σ, which takes $2 \to 3$ and $3 \to 2$, leaving 1 unaltered. It satisfies $\sigma^2 = E$. The transformation can be represented by the 3×3 matrix in

$$\begin{pmatrix} V_1 \\ V_2 \\ V_3 \end{pmatrix} \to \begin{pmatrix} V_1 \\ V_3 \\ V_2 \end{pmatrix} = \begin{pmatrix} 1 & 0 & 0 \\ 0 & 0 & 1 \\ 0 & 1 & 0 \end{pmatrix} \begin{pmatrix} V_1 \\ V_2 \\ V_3 \end{pmatrix}. \tag{2.2}$$

To obtain the effect of two operations, R followed by σ, all that is necessary it to take the matrix product, σR, where σ and R are the 3×3 matrices that represent the rotations. (It may be noticed that σR is not the same as $R\sigma$.) This example has shown that, by introducing a labeling, rotations which leave an unlabeled equilateral triangle invariant can be represented by 3×3 matrices. Similarly, the rotations that leave an unlabeled cube invariant can be represented by 8×8 matrices. Since any such matrix raised to the appropriate power will equal a unit matrix, it follows that the determinant of any such matrix is either 1 or -1.

The next question, for a cube, is: How many different matrices are there? This can be established in a variety of ways. Suppose that the vertex labeled 1 is at (a, a, a). Then a rotation through $2\pi/3$ about the 1-8 axis leaves the cube invariant but gives a rearrangement of some of the labels, as does a rotation through $4\pi/3$. These three arrangements are different from any other three that have 2, 3, etc., at (a, a, a), so altogether there are twenty-four different arrangements of the labels. This means that there will be twenty-three 8×8 different matrices that describe changes from an initial choice, and a further matrix, an 8×8 unit matrix, which it is convenient to include in the set, which is the "rotation" which leaves the labeling unaltered (a rotation through an angle that is zero, or a multiple of 2π). Once these twenty-four matrices have been obtained, any sequence of rotations that leave the cube invariant can be simulated by a product, reading from right to left, of the corresponding matrices. The set of twenty-four matrices is called a "representation" of the rotations that leave a cube invariant. It forms a group known as the cubic group.

The properties that allow the set of matrices to be described as a group are as follows. All the matrices have to be different and the product of any two, in either order, must produce a matrix that already belongs to the set. Each matrix must also have a unique inverse, which also belongs to the set, and is such that the product of the matrix and its inverse, in either order, gives a specific matrix, the unit matrix. For an abstract group, such as the group of rotations that leave the cube invariant, there is a similar set of requirements. The unit element is then usually denoted by E and the inverse of an element, A, is denoted by A^{-1}. It is such that $A^{-1}A = AA^{-1} = E$. In the context of rotations of a cube, A could be a rotation about a $(1, 1, 1)$ axis through $2\pi/3$. A^{-1} would then be a rotation about the same axis through $-2\pi/3$. Matrices automatically have an associative property, that $(AB)C = A(BC)$: a similar requirement is imposed on the elements of abstract groups. Such a group can be defined solely by its multiplication table. That is, for every pair of elements, say A and B, it is necessary to know which element is equal to AB and which to BA. For matrix groups it is seldom necessary to write out the matrices, for, like abstract groups, their properties can be deduced from knowing the multiplicative properties of all pairs of elements. These make the group elements unique, for if AB and AC both gave D then applying A^{-1} to the left of each pair would make $B = C$ and the set would not form a group, for B and C would not be distinct elements.

The example of the cube has shown that rotations can be represented by matrices, and the usual practice is to use the same notation for a rotation and for the matrix that it produces. While the two are clearly not identical, there is sufficient similarity for them to be regarded, for some purposes, as "identical." The term "isomorphic" has therefore been coined to describe this special concept of "identical." The set of rotations that leave a cube invariant is said to form a group which is "isomorphic" with the above 8×8 matrix group, for both groups have the same number of elements and the same multiplication table, provided, of course, that the elements in the two groups are labeled so as to correspond to one another. (A cube has fourfold as well as threefold axes of rotation. It would be unsatisfactory to label an element that rotates the cube by $2\pi/3$ about a threefold axis by A, and so satisfies $A^3 = E$, and an element in the matrix group also by A if it has the property that $A^4 = E$, which would describe a $\pi/2$ rotation about one of the coordinate axes.)

§ 2.2 Isomorphisms

The concept of isomorphism is extremely useful, for it can be used to relate the symmetry properties of what might otherwise be regarded as quite different systems. For example, the six points that are the centers of the faces of the above cube form a regular octahedron, and it is immediately apparent that the rotations which leave the cube invariant also leave the octahedron invariant. The symmetry group of the octahedron is therefore isomorphic with that of the cube. But the octahedron has only six vertices, so if these are numbered from 1 to 6 a treatment along the lines given for the cube will produce a set of twenty-four 6×6 matrices, which in one sense is another group. But as far as symmetry properties are concerned the group of rotations of the cube, the set of twenty-four 8×8 matrices it generates, the group of rotations of the octahedron and the set of twenty-four 6×6 matrices it generates are all identical. The groups are said to be isomorphic, though the individual members are only isomorphically related if they are appropriately ordered.

There are many different families of isomorphic groups, and group theory is concerned with the properties common to a given family and those which may exist between families. That there can be relationships between different families can be seen by considering what happens to the cube if it is compressed along one of its body diagonals. This lowers the symmetry, without necessarily eliminating it completely, for the axis of compression may continue to be a threefold axis of rotation. The rotations that leave the distorted cube invariant will form a group, but one with much fewer elements than the cubic group. Nevertheless all its elements will be elements in the cubic group, so the new group is regarded as a subgroup of the cubic group. Another group, related to the cubic group, is obtained by considering the regular tetrahedron formed by taking every other vertex of the cube. For some of the cubic rotations it will be invariant, while for others it will change into a different tetrahedron.

The fact that apparently different groups can be isomorphic does not mean that this is always obvious, so there is often an interest, when some new group is introduced, in knowing whether or not it is isomorphic with some family of groups that is already well known. A useful first step is to have a technique that will convert any group with a finite number of elements into an isomorphic matrix group. To do this the elements E, A, B, etc., are set out in some convenient order. A particular element, X, is then chosen and each element is premultiplied by it, to produce the sequence XE, XA, XB, etc. From the properties of a group each member of the sequence is a member of the group and no two elements are the same, for if XA happened to equal XB an application of X^{-1} would show that $A = B$, which is not allowed. Thus the effect of X on the sequence of elements is analogous to the effect of a cubic rotation on the vertices of a cube: it reorders the sequence. So X, and therefore every element of the group, can be represented by a unique $N \times N$ matrix, where N is the number of elements in the group. The set of matrices forms a group that is isomorphic with the original group and is known as the "regular representation" of the initial group. It is not unique, however, for had the procedure started with the group elements arranged in some other order, say B, E, D, etc., the matrix obtained for X would have been different. On looking at the matrices of two such regular representations, it would probably not be obvious that they are isomorphic, though the relationship between the two is easily found.

The sequence B, E, D, ... can be turned into the sequence E, A, B, ... by writing them both as column matrices and defining a square matrix, T, consisting of elements which are either 0 or 1, which converts one column matrix into the other. The reordering of the sequence can be written as $Tu = v$, where u is the column matrix that represents the first arrangement and v is the column matrix for the reordered sequence. Suppose now that the effect of a group element, R, on the initial sequence, u, is represented by a matrix R. Then its effect on u is to replace it by Ru. Its effect on v, which is equal to Tu, would be to replace v by TRu. But $u = T^{-1}v$, so the overall effect on v would be to replace it by $TRT^{-1}v$. Thus the matrix of the operation, R, when it is applied to the u sequence, is equivalent to applying the matrix TRT^{-1} to the v sequence. As R can be any element of the group and $(TAT^{-1})(TBT^{-1}) = TABT^{-1}$, it follows that the multiplication table of the representation based on u is the same as that based on v. (It may be noted that while the matrix T may be an element of the matrix group, this is not necessary. Any matrix that has an inverse will do.)

Returning to the cubic group, its regular representations will consist of twenty-four 24×24 matrices. They will, though, contain no more information about the structure of the group than the set of twenty-four 8×8 matrices or the set of twenty-four 6×6 matrices obtained by inspection of the effects of various rotations. So if it comes to making a choice of which representation to use in trying to find out more about the group properties, there seems little doubt that the 6×6 set, or some set of smaller matrices, if it can be found, would seem to be more convenient. The concept of a regular representation should not, though, be readily dismissed, for it has one overriding property, that it can be found for any finite group. For this reason it plays an important role in the establishment of a number of general properties of group representations. Nevertheless, it is the properties, rather than how they are proved, which really matter, so from now on little attention will be paid to regular representations, and such theorems as are needed will simply be quoted. (References to these and other group properties can be found at the end of this chapter.)

So far, the representations have been sets of matrices in which all the matrices have been different. Each representation has therefore been a group, as well as being a representation of a group. It may therefore be supposed that there is no distinction between the two and that all matrix representations form groups. In fact, there is a difference, for the representation is composed of matrices whereas the group it represents may consist of quite different entities—rotations, for example!

It is easy to demonstrate that matrix representations do not have to be groups. Suppose that instead of labeling the three coordinate axes of a cube by Ox, Oy, and Oz they had been given different colors, such as red for Ox, white for Oy, and blue for Oz. Then any rotation that leaves the cube invariant has the potential to rearrange the coloring and so it can be represented by a 3×3 matrix. However, the twenty-four matrices so obtained will not all be different, for a rotation through π about any one of the three axes will leave the color scheme unaltered. So at least three different rotations will be represented by the same 3×3 matrix, which is enough to show that this representation of the cubic group does not form a group. It nevertheless has the same multiplication table, for if two successive rotations A and B are equivalent to a rotation C, so that $BA = C$, then the 3×3 matrices derived for each of them using the color scheme, which for clarity will be written as a, b, and c, satisfy $ba = c$.

An extreme, but important, case of a representation that is not a group is obtained when every element is represented by the same unit matrix. If elements A and B are related to C by $BA = C$, then on setting $a = b = c = I$, a unit matrix, it is immediately apparent that $ba = c$. This shows that every finite group has a set of representations in common, the $n \times n$ unit matrices, where n is any integer. Choosing different values for n is, though, an unnecessary elaboration, so the convention is to restrict it to unity. This representation, which is common to every finite group, is often denoted by A_1.

In addition to the rotations that leave a geometrical shape invariant, there is often another symmetry operation which also does this, inversion in the origin $(x \rightarrow -x, y \rightarrow -y, z \rightarrow -z)$. In the cubic group it amounts to interchanging the integers on opposite ends of the body diagonals, an operation which produces an ordering of the numbering that cannot be matched by any rotation of the original ordering. This is easily seen by using the property that inversion can be combined with $(x \rightarrow -x, y \rightarrow -y, z \rightarrow z)$, a rotation through π about Oz, to give $(x \rightarrow x, y \rightarrow y, z \rightarrow -z)$, a reflection in the Oz plane. Including inversion extends the cubic group to one with forty-eight elements. Inversion also interchanges the two tetrahedra that can be imbedded in a cube.

§ 2.3 Bases for Representations

There is another way of looking at the cubic group, for instead of rotating the cube holding the coordinate axes fixed there is the alternative of holding the cube fixed and rotating the coordinate axes. Among the possible rotations there is a rotation through $2\pi/3$ about the $(1, 1, 1)$ axis, which changes Ox into Oy, Oy into Oz, and Oz into Ox. Another is a rotation that changes Ox into Oy and Oy into $-Ox$, while leaving Oz unchanged, and so on. Such changes introduce a new feature, the possibility of generating representations using algebraic expressions.

Consider, for example, the position coordinates (x, y, z) of an arbitrary fixed point and how these change when the coordinate axes are rotated so as to leave a cube invariant. A rotation about Oz through $\pi/2$ will change $\{x, y, z\}$ into $\{y, -x, z\}$, where $\{\ldots\}$ is used to denote a sequence, an ordered arrangement of algebraic quantities. The effect on the sequence can be represented by the 3×3 matrix in

$$
\begin{pmatrix} y \\ -x \\ z \end{pmatrix} = \begin{pmatrix} 0 & 1 & 0 \\ -1 & 0 & 0 \\ 0 & 0 & 0 \end{pmatrix} \begin{pmatrix} x \\ y \\ z \end{pmatrix}, \tag{2.3}
$$

which shows that a new feature, a change in sign of an element in the sequence, is readily incorporated in a matrix representation. The sequence $\{x, y, z\}$ can therefore be used to obtain a 3×3 representation of the cubic group. Another example uses the sequence $\{xy, yz, zx\}$ to obtain a 3×3 representation, raising the question of whether or not it is the same as the one obtained with $\{x, y, z\}$.

Before answering this question, it is necessary to decide what is meant by "different," for it has already been shown that a reordering of the elements in a sequence changes the matrices in the representation. As such a change replaces A by TAT^{-1}, there is one quantity that stays unaltered, the sum of the diagonal elements, which is known in group theory as the character of the matrix. Another is their multiplication table. With algebraic expressions in the sequence there is much more scope for change, such as changing $\{x, y, z\}$ into $\{3x, 2y, z\}$ or into $\{(x + y), (x - y), z\}$, or even into $\{(x + iy), (x - iy), 2z\}$.

In each case the change can be represented by operating on a column matrix by a suitably chosen square matrix, T, the elements of which do not depend on the values which may be given to x, y, and z. For consistency T should have an inverse, so that the two sequences are uniquely related. The representations that result from such changes have quite different looking matrices for a given matrix element, A say, but even so all the matrices that represent A have the same character, and furthermore the same multiplication table. Thus, as far as these two properties are concerned, two different sequences will give identical properties provided they are related by a linear transformation T that has an inverse. It is therefore convenient to generalize the concept of a "sequence" to that of a "basis," the property of which is to give a matrix representation of the group that correctly reproduces its multiplicative table and to give a unique character to the matrix that represents a particular group element. Two representations are therefore only regarded as different if they have different characters for some of the group elements (they will not be representations if they do not reproduce the multiplication table).

Returning to the $\{xy, yz, zx\}$ basis for the cubic group, it is instructive to replace xy by xyz/z, yz by xyz/x, and zx by xyz/y and then reorder the basis into $\{xyz/x, xyz/y, xyz/z\}$, a step which, with the above definition of change of representation, is not regarded as a change in basis. In the new form it can be seen that all the basis elements now have the same numerator and that this either remains unaltered or changes its sign under the operations of the group, while the denominators change in exactly the same way in this basis as they would in the $\{x, y, z\}$ basis. Thus, on comparing the matrices obtained with the $\{x, y, z\}$ basis with those from the $\{xyz/x, xyz/y, xyz/z\}$ basis, it can be seen that some of them will be identical and others will have elements that are reversed in sign, according to whether xyz goes into xyz or into $-xyz$. So the two sets of matrices will be rather similar, but whether they are to be regarded as the same or different is determined by the sums of the diagonal elements and not by the individual matrix elements. The operation $x \to y$, $y \to -x$, $z \to z$, a rotation through $\pi/2$ about Oz, takes $xy \to -yx$, $yz \to -xz$, $zx \to -zy$, and $xyz \to -yxz$, or

$$\begin{pmatrix} -yx \\ -xz \\ -zy \end{pmatrix} = \begin{pmatrix} -1 & 0 & 0 \\ 0 & 0 & -1 \\ 0 & -1 & 0 \end{pmatrix} \begin{pmatrix} xy \\ yz \\ zx \end{pmatrix}, \tag{2.4}$$

and the character of the matrix is -1, whereas the character in the $\{x, y, z\}$ basis is $+1$. [All the elements in the matrix in (2.4) are reversed but as there is only one nonzero element on the diagonal it is only this reversal that matters in determining the character.] The two 3×3 representations are therefore different.

Another possibility is to use $\{x^2, y^2, z^2\}$ as a basis for a 3×3 matrix representation. However, this basis can be regarded as the same as that which uses

$\{x^2 + y^2 + z^2, x^2 - y^2, z^2\}$. It has the interesting property that the application of any group operator to $x^2 + y^2 + z^2$ leaves its form unaltered, so the first row and first column of each matrix will have the same sequence of elements: $1, 0, 0$. It would therefore have been neater to have separated $x^2 + y^2 + z^2$ from the other two elements in the basis, for it alone will produce a 1×1 matrix representation, which is recognizable as the unit representation. Having separated $x^2 + y^2 + z^2$, there will be transformations on $x^2 - y^2$ and z^2 that will form a basis for a 2×2 representation. [A frequent alternative choice is $\{(x^2 - y^2), (3z^2 - r^2)\}$, for $(y^2 - z^2)$ and $(z^2 - x^2)$ can be written as linear combinations of them.] This example shows that while an initial choice gave a 3×3 representation, a change in basis made it possible to regard it as providing a 1×1 and a 2×2 representation. Two questions then arise. Is it possible to reduce the 2×2 representation to two 1×1 representations, by some further rearrangement of the basis, and can the other 3×3 representations based on $\{x, y, z\}$ and $\{xy, yz, zx\}$ be similarly reduced?

§ 2.4 General Properties of Irreducible Representations

At this point it is convenient to introduce the concept of a "block diagonal" matrix. Matrices do not need to have the same number of rows and columns so it is possible to have matrices of matrices, a simple example being

$$\begin{pmatrix} M_{n,n} & M_{n,m} \\ M_{m,n} & M_{m,m} \end{pmatrix}, \tag{2.5}$$

where $M_{n,n}$ is an $n \times n$ matrix, $M_{n,m}$ is $n \times m$, $M_{m,n}$ is $m \times n$ and $M_{m,m}$ is $m \times m$. The interest now is in examples where $M_{n,m}$ and $M_{m,n}$ are null matrices so that the whole matrix is "block diagonal." For if, in a matrix representation of a group, all the matrices have the same block diagonal form, then further matrix representations can be obtained by using either just the $M_{n,n}$ parts or just the $M_{m,m}$ parts of each. This is what happened, for the cubic group, on changing the $\{x^2, y^2, z^2\}$ basis to $\{x^2 + y^2 + z^2, x^2 - y^2, 3z^2 - r^2\}$. It resulted in every matrix in the 3×3 representation taking an identical block diagonal form, a 1×1 block and a 2×2 block. The 1×1 blocks and the 2×2 blocks give separate representations of the group.

The possibility of reducing the 2×2 representation into two 1×1 representations will not be explored here, but had it been it would have been found to be impossible. Both the 1×1 and the 2×2 representations are examples of "irreducible representations," representations for which it is not possible to change the basis so that each matrix decomposes into a two or more block diagonal form. (The abbreviation i.r. is commonly used for "irreducible representation.") For general $(n + m) \times (n + m)$ block diagonal matrices, there is always the possibility that the $(n \times n)$- and/or the $(m \times m)$-dimensional representations can be further reduced. The process of transforming a basis so as to reduce all the matrices in a representation to a common block diagonal form that cannot be further reduced is described as reducing the representation to a sum of irreducible representations. It is often written as

$$\Gamma = \Gamma_i + \Gamma_j + \Gamma_k + \cdots,$$

where the left-hand side represents the reducible representation and the right-hand side represents it in terms of irreducible representations, the Γ_i's, etc., blocks. They can be regarded as the diagonal elements of a matrix of matrices in which all the off-diagonal elements (which are also matrices) are null. The order in which the Γ_i's are written is arbitrary. Within a given representation and therefore within each i.r. a change of basis can change the matrices without changing the characters or the multiplication table. Since the character of a matrix in a reducible representation is unchanged when the basis is altered, it follows that the character of each matrix of a reducible representation is the sum of the characters in its expression as a sum of i.r.'s.

The concept of the characters of representations and therefore the characters of i.r.'s can be further simplified because usually many of the matrices in a representation have the same character. Suppose an arbitrary element, X, is used to rearrange the order of the elements of a group, by using it to change each element, A say, into the element XAX^{-1}. This is either the same as A or another group element. But whichever it is, A and the element XAX^{-1} will have the same character. So it is now only a matter of running through all possible choices of X to produce a set of operators all of which will have the same character in any representation. Such a family is called a "class," and there are at least two classes in every group, for any such transformation applied to the unit operator gives the unit operator, so this operator alone forms one class. Continuing the process, it can be seen that for all groups the elements can be put into classes, so there is no need, in a representation, to list the characters of every group element; it is sufficient to list the character of a typical element in each class. This has two consequences. First, the number of classes is usually much smaller than the number of elements in the group and, second, an important theorem can be proved, that the *number* of different irreducible representations is equal to the *number* of classes. It is then sufficient to have a notation for each class, to write them in some order, and to produce a table that gives the characters of each class, for each i.r., in the same order. One i.r. is then regarded as different from another if its entry in the character table contains some different numbers.

The whole procedure can be illustrated using the cubic group, which is usually denoted by O, as an example. The group has twenty-four elements, which can be separated into five classes, the first of which contains only one element, E, the unit operator, which corresponds to leaving the cube where it is. The second, denoted by $8C_3$, contains eight elements, rotations through $2\pi/3$ and $4\pi/3$

Table 2.a. The Character Table of the Cubic Group

O		E	$8C_3$	$3C_2$	$6C_2$	$6C_4$
$(x^2 + y^2 + z^2)$,	A_1	1	1	1	1	1
(xyz),	A_2	1	1	1	-1	-1
$(x^2 - y^2, 3z^2 - r^2)$,	E	2	-1	2	0	0
(x, y, z),	T_1	3	0	-1	-1	1
(xy, yz, zx),	T_2	3	0	-1	1	-1

about the four threefold symmetry axes. (The 8 in $8C_3$ denotes the number of elements in the class, the C that they are rotations, and the 3 that each element in the class satisfies the relation $C^3 = E$.) The third, $3C_2$, contains three elements, the three rotations through π about axes through the centers of faces, the fourth, $6C_2$, contains six rotations through π about axes through the midpoints of opposite edges, and the fifth, $6C_4$, contains six rotations through $\pm\pi/2$ about axes through the centers of opposite faces. The character table, the list of the characters of the classes in the i.r.'s, is given in Table 2.a. It has been obtained using a variety of theorems about i.r.'s.

The group label, O, is given in the top left-hand cell. The rest of the top row gives the labeling of the classes and the rest of the first column gives the labeling of the irreducible representations, the A_1, A_2, etc. These are on the right, with examples of bases $(\ldots, \ldots, \ldots)$ that can be used to obtain them placed to their left, so that the labels of the i.r.'s are adjacent to their characters. The rest of the table gives the characters, which are often denoted as χ's. The class labeled E contains only one element, the unit operator. Its matrix is always a unit matrix, so the values of χ in its column give the dimensions of the matrix representation of the corresponding i.r. Thus the T_1 i.r. consists of 3×3 matrices. In interpreting the properties of the rest of such a table, it is often useful to realize that although it is not possible to reduce the block diagonal form of all the matrices in an i.r. any further, it is possible to chose a basis in which the representation of a specific group element is completely reduced to diagonal form. Thus, if the matrix of a particular group element in C_3 is reduced to a diagonal form, it can only have 1, ω, and ω^2, the cubic roots of unity, as diagonal elements because the group element satisfies $X^3 = E$ and so must its representation. Thus the -1 value for this class in the E representation, which is 2×2, shows that one of its diagonal elements will necessarily be ω and the other one will necessarily be ω^2, for it is only by adding these two that -1 can be obtained. In A_2, which is a 1×1 representation, the -1 entry for C_4 shows that its matrix element is -1, and since the square of the corresponding operator belongs to C_2 its matrix element must be 1. These are just a few examples of consistency in the table of group characters.

The character tables for the i.r.'s of all the groups likely to be met in magnetic studies have long been known. They can be found in many publications (see the end of this chapter). The first row is always 1, 1, 1, ..., and it is usually not difficult to check the other entries using elementary arguments similar to those given above, as an alternative to using the various theorems about group representations. For the cubic group, with the bases listed, a rotation belonging to the C_2 class takes x into $-x$, y into $-y$, and z into z, so giving -1 for the characters of both the T_1 and T_2 i.r.'s. For the E i.r. the same operation takes $(x^2 - y^2)$ and $(3z^2 - r^2)$ into themselves, giving 2 for the character. A slightly more complicated example is that of an operator in the C_3, class which takes x into y, y into z, and z into x. This takes $(x^2 - y^2)$ into $(y^2 - z^2)$ and $(3z^2 - r^2)$ into $(3y^2 - r^2)$, so both the new forms have then to be written as linear combinations of the original basis forms. Thus $(y^2 - z^2)$ is rewritten as $-1/2[(x^2 - y^2) + (3z^2 - r^2)]$ and $(3y^2 - r^2)$ as $-1/2[3(x^2 - y^2) + (3z^2 - r^2)]$. Both contribute $-1/2$ to the character, making its value -1.

Among the theorems about group i.r.'s there is one that shows how to decompose a reducible representation into i.r.'s, and as this is particularly useful in physics it will be given explicitly. Suppose that a given representation consists of $N \times N$ matrices. The first step is to obtain the character of an element in each class. Then the number of times, a_i, that the Γ_i irreducible representation occurs is given by the formula

$$a_i = \frac{1}{h} \sum_R \chi(R)\chi_i(R), \tag{2.6}$$

where h is the number of elements in the group, $\chi(R)$ is the character of the group element R in the $N \times N$ representation, and $\chi_i(R)$ is its character in the ith irreducible representation. The summation is over all the group elements, with all the elements in a given class giving the same contribution. Another theorem, which it is useful to know, is that if the i.r.'s of E are squared and summed over all i.r.'s, the total equals h.

The i.r.'s of the cubic group have dimensions 1, 2, and 3, so it is obvious that the 8×8 representation obtained by labeling each vertex is reducible. With the character table and the theorem given in (2.6) it can be written as a sum of i.r.'s. The matrix that represents E is the 8×8 unit matrix, so the sum of its diagonal elements, $\chi(E)$, is 8. The elements in the C_3 class contain rotations of $2\pi/3$, which leave only two vertices unchanged. So the matrices which represent the elements in this class all have the same character, $\chi(C_3)$, which is equal to 2. The elements in the C_2 class induce rotations of π about coordinate axes. No vertices are left unchanged, so $\chi(C_2)$ is 0. Similarly, all the vertices are changed with the operators in both the second C_2 and in C_4. Thus the sequence of characters in the order of the classes is 8, 2, 0, 0, 0. Using eqn. (2.1) it follows that $a_1 = 1$, $a_2 = 1$, $a_3 = 0$, $a_4 = 1$ and $a_5 = 1$. The reduction of this representation into its irreducible classes can therefore be written as

$$\Gamma = A_1 + A_2 + T_1 + T_2.$$

Another example is provided by the 6×6 representation based on the octahedron formed from the centers of the faces of the cube, the character sequence for which is 6, 0, 2, 0, 2. Its Γ reduces to $A_1 + E + T_1 + T_2$.

§ 2.5 Irreducible Representations in Physics

Irreducible representations are used a good deal in physics and at last the point has been reached at which it becomes possible to give the reader some indication of this. The first use is in labeling degeneracies, for it can usually be assumed that when operators are found to have degenerate eigenstates, families of orthogonal states which have the same eigenvalue, it is because of symmetry properties. It is then worthwhile, if it is not already obvious, to identify the symmetry group. (In many cases the symmetry will be a symmetry in operators. For example, in a cubic environment the Hamiltonian often contains a term of the form $x^4 + y^4 + z^4$, where x, y, and z are *operators*. Under these conditions the symmetry group is a group of operations, the elements of which have to be defined. They are then applied to *operators*, a point that will be given further consideration in due course.) For the moment it will be assumed that the symmetry group of the operator, which since it is often a Hamiltonian will be referred to as such, is known. Then it can be expected that on applying one of the symmetry operations to one of the eigenstates of the Hamiltonian, which itself is left unaltered by the operation, the state that results will either be a multiple of itself or a linear combination of itself and some other state, which is therefore orthogonal to it. By applying all the group operators to all the orthogonal states so generated, a finite number of orthogonal states is usually generated. These can be used to give a representation of the symmetry group of the Hamiltonian. This representation is often irreducible, in which case the notation for the appropriate i.r. can be used to label the eigenvalue and so indicate the symmetry properties of its eigenstates. Sometimes, however, the representation is reducible, in which case it should be written as a sum of irreducible representations and these should be used for the labeling of the states.

Reducible representations occur most commonly when the operator of interest is an approximation to the Hamiltonian of actual interest. It is then probable that the Hamiltonian will have either the same or a lower symmetry than that of the approximate Hamiltonian. (It is not a good practice to have an approximate Hamiltonian that has less symmetry than the Hamiltonian of interest, for this greatly complicates deriving the properties of the actual Hamiltonian

Table 2.b. The Character Table of the Symmetry Group D_4

D_4		E	C_2	$2C_4$	$2C_2'$	$2C_2''$
(z^2),	A_1	1	1	1	1	1
(z),	A_2	1	1	1	−1	−1
	B_1	1	1	−1	1	−1
	B_2	1	1	−1	−1	1
(x, y),	E	2	−2	0	0	0

from the approximate one.) If the two have the same symmetry any degeneracy associated with a reducible representation of the approximate Hamiltonian is probably accidental, in the sense that a change from the approximate to the actual Hamiltonian is likely to lift the degeneracy. So the energy level is likely to be split into a number of levels, each of which can be associated with one of the i.r.'s of the original reducible representation.

The important point here is that, for a given symmetry of the Hamiltonian (which for this purpose can be any operator) the degeneracy associated with an i.r. is a fundamental one, for there is no way in which the basis states associated with it can be separated in energy by any change that preserves the symmetry. This does not mean that the basis states will not change, for in general they will, but rather that as the Hamiltonian is changed they will change in such a way that the degeneracy is not lifted. Thus, knowing the decomposition of a reducible representation of an approximate Hamiltonian gives an indication of how the degeneracy will be split on moving to the actual Hamiltonian, *provided the symmetry remains unaltered*. If, however, the symmetry is lowered, it can be expected that the degeneracy associated with a reducible representation of the higher symmetry group will be lifted, and, additionally, the degeneracies of its i.r.'s will also be lifted. Any degeneracies that remain will most likely be associated with i.r.'s of the lower-symmetry group.

To deal with a lowering of symmetry it is necessary to have the character tables of all the common symmetry groups. As an example, it can be supposed that the symmetry of a cube has been reduced by a compression along the z axis, so that the faces perpendicular to this axis remain squares. The new symmetry group, which is usually denoted by D_4, has the character table, shown in Table 2.b. It has five classes. The first, E, contains, as usual, just the unit operator. The second, C_2, also contains just one operator, the rotation through π about the axis of compression. C_4 contains two operators that can be recognized as rotations through $\pi/2$ and $3\pi/2$ about Oz. The remaining two classes both contain two operators, rotations through π. These are the rotations through π about the Ox and Oy axes and the rotations through π about the axes which join the midpoints of opposite z-directed edges. The first pair belong to one class and the second to another. The classes are distinguished by primes. It does not matter which pair is given the single prime, for a change to two primes simply interchanges B_1 and B_2. (The interchange is also induced by a rotation of the Ox and Oy axes through $\pi/4$ about Oz, which then directs them through the centers of opposite edges rather than through the centers of opposite faces.)

With this character table it is possible to see what happens when the symmetry of the cubic group, O, is reduced to that of D_4. Suppose that when the symmetry is cubic a given energy level is spanned by a T_1 representation. All the elements of D_4 belong to O and their characters in O, when ordered in the same way as the classes of D_4 are 3, -1, -1, 1, 1. Using the character table of D_4 this sequence is that of a reducible representation, which when expressed as a sum of i.r.'s is $A_1 + E$. It can therefore be anticipated that the threefold T_1 degeneracy in cubic symmetry will be split, when the symmetry is lowered by the compression, into a singlet and a doublet, spanned respectively by the A_1 and the E i.r.'s of D_4.

Group theory does not usually give magnitudes or signs for splittings; in the above example, it has been used to show how a degeneracy may be lifted by a lowering of symmetry and the symmetry properties of the degeneracies that will be left. Such information is often worth having.

In quantum mechanics the eigenstates and eigenvalues of the Hamiltonians of interest can seldom be determined exactly. There is therefore a good deal of interest in approximate Hamiltonians that have the same or higher symmetries and for which the eigenvalues, the eigenstates, and the associated i.r.'s can be found. The behavior to be expected on symmetry arguments can then be followed from one to the other by setting $H = H_0 + \lambda(H - H_0)$, where H is the actual and H_0 is the approximate Hamiltonian, and letting λ increase from zero to unity.

Other uses of i.r.'s include using them to show that specific matrix elements are zero and that the nonzero matrix elements of an operator of interest can be arranged into families such that all those in a particular family are proportional to those of another, symmetry-related operator, the matrix elements of which are easy to find. The operator of interest can thus be written in a parametrized form, an expression in familiar operators each of which is multiplied by an undetermined parameter. At this stage it is not easy to find simple examples to illustrate this, though a number will be given in due course, particularly in connection with crystal field theory.

§ 2.6 Continuous Groups

So far it has been convenient to concentrate on groups that have a finite number of elements. Now the concept of continuous groups will be introduced, using the rotations of axes that leave a sphere invariant. If a point is given coordinates x, y, z and the axes are rotated so that it becomes X, Y, Z, the two descriptions are related thus:

$$\begin{pmatrix} X \\ Y \\ Z \end{pmatrix} = \begin{pmatrix} R_{11} & R_{12} & R_{13} \\ R_{21} & R_{22} & R_{23} \\ R_{31} & R_{32} & R_{33} \end{pmatrix} \begin{pmatrix} x \\ y \\ z \end{pmatrix}, \qquad (2.7)$$

where the 3×3 matrix is an orthogonal matrix which can be used to represent the rotation. The invariance is that associated with the substitution operation:

$$\begin{pmatrix} x \\ y \\ z \end{pmatrix} \rightarrow \begin{pmatrix} R_{11} & R_{12} & R_{13} \\ R_{21} & R_{22} & R_{23} \\ R_{31} & R_{32} & R_{33} \end{pmatrix} \begin{pmatrix} x \\ y \\ z \end{pmatrix}. \qquad (2.8)$$

Since there is different 3×3 matrix for each different rotation, it follows that the rotation group can be represented by an infinite set of 3×3 matrices. Also, as it is not possible to find a (T^{-1}, T) transformation which, acting on the matrices, will reduce them all to a common block diagonal form, the representation is irreducible. As for the finite groups the elements can be arranged into classes and, with an infinite number of group elements, it is not surprising that the number of classes and the number of i.r.'s are also infinite.

A useful way of studying the i.r.'s is provided by the solutions of Laplace's equation

$$\frac{\partial^2 V}{\partial x^2} + \frac{\partial^2 V}{\partial y^2} + \frac{\partial^2 V}{\partial z^2} = 0, \qquad (2.9)$$

which provides a simple example of a partial differential equation that is rotationally invariant. On changing to new coordinates OX, OY, and OZ it becomes

$$\frac{\partial^2 V}{\partial X^2} + \frac{\partial^2 V}{\partial Y^2} + \frac{\partial^2 V}{\partial Z^2} = 0, \qquad (2.10)$$

and the invariance appears when the substitution in (2.8) is used in (2.9). If the differential equation is now expressed in spherical polar coordinates, its solutions can be written as products of powers of r and angular functions, the associated Legendre functions, thus:

$$V = r^k P_l^m (\cos \theta) e^{im\phi}, \qquad (2.11)$$

where $k = l$ or $-(l + 1)$ and $r^2 = x^2 + y^2 + z^2$. Since r is invariant and the spatial dimensions of any solution are not changed by a rotation, it is clear that any substitution applied to a solution with a particular value for l produces a linear combination of solutions all with that same l. In other words, the solutions for a given l can be used to generate a $(2l + 1) \times (2l + 1)$ matrix representation of the rotation group, there being $(2l + 1)$ different values for m for a given value of l. Since many of the basis elements are complex, many of the matrix elements will also be. Such a representation is irreducible. It is usually denoted by D_l. The x, y, and z solutions provide a basis for the D_1 i.r.

From this basis nine other functions can be obtained by taking the components of two vectors (x_1, y_1, z_1) and (x_2, y_2, z_2) and multiplying each component of one by all the components of the other. This produces nine functions: x_1x_2, y_1y_2, z_1z_2, x_1y_2, y_1x_2, etc. They can be used as a basis for a 9×9 matrix representation of the rotation group, for they clearly transform into linear combinations of one another under rotations. This representation is reducible, for the linear combination $x_1x_2 + y_1y_2 + z_1z_2$ is the scalar product of the two vectors and so is an invariant under all rotations. It can therefore be used as a basis for the D_0 i.r. It can also be seen that $(y_1z_2 - z_1y_2)$, $(z_1x_2 - x_1z_2)$, and $(x_1y_2 - y_1x_2)$, which are the components of the vector product of the two operators, have exactly the same transformation properties as x, y, and z, a basis set for a D_1 i.r. They can therefore be used as another basis set for the D_1 representation. What is not quite so obvious is that the remaining five functions give a D_2 i.r. That this is so can be seen by comparing the set $\{(x_1y_2 + x_2y_1), (y_1z_2 + y_2z_1), \ldots\}$ with the set $\{xy, yz, zx, (x^2 - y^2), (3z^2 - r^2)\}$ (in the last of which r^2 is replaced by $x_1x_2 + y_1y_2 + z_1z_2$), all of which are solutions of Laplace's equation that are quadratic in x, y, and z and so form a basis for the D_2 i.r. It follows that the representation by 9×9 matrices can be written as a sum of i.r.'s, $D_0 + D_1 + D_2$. This relation is usually written as

$$D_1 \times D_1 = D_0 + D_1 + D_2. \tag{2.12}$$

As for the finite groups, it makes no difference to the characters if the basis elements are scaled or linearly transformed by a nonsingular T. It is usually convenient, however, to choose the basis so that the matrices are unitary, and to do this the P_l^m are replaced by the Y_l^m functions [see eqn. (1.8)].

The above resolution of a representation based on a product of the functions in two sets each of which formed an i.r. was done by elementary means. It could have been done using a general result for the rotation group, which gives the decomposition of the outer product, as it is called, of a basis obtained by multiplying together the basis elements of two i.r.'s, D_p and D_q. The required formula, which can be derived in a variety of ways, is

$$D_p \times D_q = D_{p+q} + D_{p+q-1} + D_{p+q-2} + \cdots + D_{|p-q|}. \tag{2.13}$$

As the relation can easily be derived using the properties of angular momentum eigenstates, the proof will be left until these have been described.

§ 2.7 The Rotation Group in Quantum Mechanics

The properties of the i.r.'s of the rotation group are used in quantum mechanics in a variety of ways, one important use being the study of angular momentum, which crops up with a variety of notations, **l** for the orbital angular momentum of a single electron, and **L** for the orbital angular momentum of many electrons (both in units of $\hbar$) being just two examples. There is also the angular momentum associated with spin, denoted **s** for a single electron and **S** for many electrons. Then **j** is used for the total angular momentum, spin plus orbit, of a single electron and **J** for many electrons. When nuclei have spins, which are usually denoted by **I**'s and combined with **J**'s, they give **F**'s, and so on. For the moment, though, it will be convenient to ignore any spin angular momentum, for it needs a separate discussion.

The symmetry properties associated with angular momentum do not depend on the notation used to describe it, so both single- and many-electron orbital angular momentum variables can be considered together, for the two descriptions are isomorphic. They all have the commutation rule $\mathbf{l} \wedge \mathbf{l} = i\mathbf{l}$, with the magnitude of **l** (in units of $\hbar$) being given by $\mathbf{l} \cdot \mathbf{l} = l(l+1)$, where l is an integer or zero. Within a manifold of given **l** the eigenstates of l_z are usually written as $|m\rangle$, where m takes all integer values between l and $-l$. The Hamiltonian of a particle moving under a central potential has spherical symmetry, and its eigenfunctions can be written as radial functions times angular functions. The usual choice for the latter is the Y_l^m functions. Then not only is there an isomorphism between the ket vectors that are the eigenstates of l_z and the wave-mechanical Y_l^m orbital wave functions, but there is also an isomorphism in many other properties. For example, $l_+ |m\rangle = [l(l+1) - m(m+1)]^{1/2} |m+1\rangle$. If l_+ is written as a differential operator, as in wave mechanics, an identical relation is obtained if $|m\rangle$ is replaced by Y_l^m and $|m+1\rangle$ by Y_l^{m+1}. Now the Y_l^m, for a definite value of l, form a basis for a D_l i.r. of the rotation group, so it follows that the $|m\rangle$ kets and the $\langle m|$ bras, for a given l, also provide bases for the D_l i.r.

The proof of eqn. (2.7) can now be readily obtained. In considering a rotation about an arbitrary axis it is simplest to take its direction as the axis of quantization and use a basis $|l, m\rangle$ in which m ranges from $-l$ to l to generate D_l. A rotation through an angle α multiplies $|l, m\rangle$ by $\exp(im\alpha)$, so the character, $\chi(\alpha)$, of the associated rotation operator is the sum of these exponential factors over all m. So

$$\chi(\alpha) = \frac{\sin \left[\frac{(l+1)\alpha}{2} \right]}{\sin \frac{\alpha}{2}} \tag{2.14}$$

is the character of any rotation through α in a D_l representation. For a product representation, such as $D_i \times D_j$, the character of the same rotation is the product of two expressions of the form of (2.14), one with $l = i$ and the other with $l = j$. (In a product representation, $\Gamma_i \times \Gamma_j$, the character of any operator, R, is the product of its characters in Γ_i and Γ_j. It will be left to the reader to show that this leads to the relation in eqn. (2.13) (with the hint that it is helpful to go back to the summations, which give the characters).

An important extension can now be made, for on physical grounds it can be expected that the Hamiltonian of any system that has electrons moving under a central potential (a potential depending only on distance from an origin), which also interact with each other (such as via Coulomb interactions, $e^2/|\mathbf{r}_i - \mathbf{r}_j|$), will be rotationally invariant (an assumption that there are no preferred directions in space). Its eigenfunctions, even though they are unknown, can then be expected to provide bases for irreducible representations of the rotation group and so be open to the D_l classification, where l can now be regarded as an angular momentum. This time, though, a capital letter would be used, because it is a many-electron system, and the i.r.'s would be written as D_L's. It is quite remarkable how many useful properties can be deduced from simply knowing this! One example is the relation

$$L_{\pm}|L, M\rangle = [L(L + 1) - M(M \pm 1)]^{\frac{1}{2}}|L, M \pm 1\rangle, \qquad (2.15)$$

which is analogous to the one for $l_{\pm}$ acting on a one-electron state, $|l, m\rangle$. Once it is recognized that the symmetry properties of the $|L, M\rangle$ basis set determine a specific D_L, which is identical with that determined by the $|l, m\rangle$ basis set when l is identified with L and m with M, the above relation follows immediately.

§ 2.8 Spin Angular Momentum

In the above discussion an assumption has been slipped in which is easy to overlook. It relates to the definition of a rotation, and thus to the meaning of a rotation group, which was initially introduced by a consideration of the rotations of a cube. In the examples, x, y, and z have been regarded as the coordinates of a point in space, whereas in quantum mechanics they are operators.

That there is a need to redefine the operators that define the groups only really becomes apparent when electron spin is introduced. Before this, it had been assumed that on rotating the coordinate axes through 2π, about any axis, the wave functions, bras and kets, would return to their original values. While this is indeed the case for the Y_l^m functions, it is not a strict requirement of wave mechanics, for the physical requirement is that it is the square modulus of a wave function which should have the 2π periodicity. When it became necessary to incorporate electron spin in the bras and kets, it was realized that the assumption of 2π periodicity was incorrect, and that it should be changed to a 4π periodicity.

The spin rotational properties follow as a consequence of an obvious extension of the concept of a rotation operator. For orbital variables the rotation operator for a single particle can be written in the form $\exp[i\alpha(pl_x + ql_y + rl_z)]$, where (p, q, r) are the direction cosines of the axis of rotation and α is the angle of rotation. The exponential is defined in terms of the power series expansion:

$$\exp[i\alpha(pl_x + ql_y + rl_z)] = \sum_{n=0}^{\infty} \frac{[i\alpha(pl_x + ql_y + rl_z)]^n}{n!}. \tag{2.16}$$

This definition is consistent with the Y_{lm} properties of the hydrogenic wave functions and the properties of the $|l, m\rangle$ kets. For a many-electron system, $\mathbf{l}$ is replaced by $\mathbf{L}$, and for spin, $\mathbf{l}$ is replaced by $\mathbf{s}$. To demonstrate how the spin kets are changed by rotations it is simplest to choose the axis of rotation as the axis of quantization. One spin state can then be written as $|+\rangle$ and the other as $|-\rangle$. A rotation through α changes $|+\rangle$ into $\exp(i\alpha s_z)|+\rangle$, or $\exp(i\alpha/2)|+\rangle$, so if $\alpha = 2\pi$ the ket is reversed in sign, and similarly for $|-\rangle$. (For a single-electron system where both the orbital and spin angular momentum are coupled, $\mathbf{s}$ is replaced by $\mathbf{j}$, and for a many-electron system $\mathbf{L}$ is replaced by $\mathbf{J}$. With an odd number of electrons there is the 4π periodicity, so the discussion of the new group will be given for this case.)

A basis for a representation of the new rotation group is obtained by using the eigenstates for a definite j, the states $|m_j\rangle$, where m_j takes values with integer spacings from j to $-j$. A rotation ϕ applied to $|m_j\rangle$ change it into $\exp(im_j\phi)|m_j\rangle$, so the calculation of the character follows the same course as is used to obtain eqn. (2.14), with j replacing l. This representation is denoted D_j. It then follows that the outer product of a D_l set of orbital states with a $D_{1/2}$ set of spin-1/2 states gives $D_{l+1/2} + D_{l-1/2}$, and that all the similar relations obtained for integer values of l, L, s, and S can be regarded as valid whether they are for integer or integer plus one-half values of momentum.

Turning to the cubic group the new definition of a rotation operator presents a problem, as can be seen by considering a rotation, A, through π, about Oz. In the geometrical picture of a cube this operator satisfies $A^2 = E$. But when what appears to be the same operator is written in the form of eqn. (2.16), it does not necessarily satisfy this relation, for the only safe assumption is that E will equal A^4. The problem arises because an expression such as $x^4 + y^4 + z^4$ is invariant if (x, y, z) are interpreted as the classical coordinates of a particle. In quantum mechanics such an interpretation is not possible, and about the best that can be done is to claim that there is a wave function for the particle which is concentrated near a vertex of a cube, and that there will be similar wave functions near the other vertices, which can be obtained from the first by applying the rotation operators of the cubic group. If "wave function" is interpreted as an orbital function then a rotation through 2π leaves it invariant. If, however, the particle is an electron then "wave function" may also include a description of its spin. If so a rotation through 2π reverses the wave function.

In the context of group theory the needed change is effected by adding an extra operator, R, to the original group, with the assumptions that it commutes with all the twenty-four previous operators and that any operator, A, previously raised to the nth power to describe a rotation through 2π, now satisfies $A^n = R$, where $R^2 = E$. If E, A, B, ... are the elements of the original group, the new group has twice as many elements, E, A, B, ..., T, TA, TB, ... The extension ensures that if $AB = C$ in the first group the same relation holds in the new group, and that $TATB$ also equals C. The new group is known as the cubic double group and denoted O^+. A similar procedure can be used to produce a double group from any group describing a spacial symmetry. In each such group there is at least one extra class, consisting solely of R, so there is at least one extra i.r., showing that double groups are different from the groups from which they are derived. [Some of the i.r.'s are similar in structure to those of the single group, though they are extended because there are more classes. The "extra" i.r.'s have characters that can usually be understood by applying suitable operators from each class to the m_j states of $j = 1/2$ and $3/2$, using the relation given in eqn. (2.14).]

Appendix B of (A-B) is a good source for finding details about the i.r.'s of the double groups, information which was particularly important in the early days of crystal field theory. The method of equivalent operators has now largely eliminated this need.

§ 2.9 Operator Basis Sets

The basis sets that have so far been chosen have included sequences of integers obtained by labeling the corners of a cube, sets of functions such as xy, yz, and zx, and eigenstates of angular momentum operators. It is evident that a wide choice is available. Now another choice will be introduced, based on noncommuting operators. The operators L_x, L_y, and L_z are analogous to x, y, and z as far as rotations are concerned, but they differ in two other respects. They do not commute and they behave differently under inversion, for the operator which takes $x \rightarrow -x$, $y \rightarrow -y$, and $z \rightarrow -z$, leaves the components of angular momentum unaltered. (The expression pseudovector is often used to distinguish $\mathbf{l}$, $\mathbf{L}$, $\mathbf{s}$, $\mathbf{j}$, and $\mathbf{J}$, which do not reverse under inversion, from vectors like $\mathbf{r}$ that do.) The inversion properties are not of particular interest at this point. They have been mentioned partly for completeness and because many of the point groups contain the inversion operator. The cubic group, O, for example, is extended when the inversion operator is included, and the new group is denoted by O_h. It contains forty-eight elements, as does the double group based on O. The two should not be identified.

Since the operators L_x, L_y, and L_z transform under rotations in the same way as do x, y, and z, they too can be used as a basis for a D_1 i.r., their noncommutative properties being of no significance. But when it comes to constructing a D_2 i.r. from them there is a difference, for whereas xy, yz, zx,

$x^2 - y^2$, and $3z^2 - r^2$ form a basis, there is bound to be some question of how they should be replaced. For example, should xy be replaced by L_xL_y, or should it be replaced by L_yL_x, or by some other expression? In fact the answer is obvious enough: the replacements should be symmetrized, for this eliminates problems over noncommutation. For D_2 an obvious basis is the set $\{(L_xL_y + L_yL_x), (L_yL_z + L_zL_y), (L_zL_x + L_xL_z), (L_x^2 - L_y^2), [3L_z^2 - L(L + 1)]\}$, with the only perhaps unexpected feature, being that, in the last member, having replaced r^2 by $L_x^2 + L_y^2 + L_z^2$, it has then been replaced by $L(L + 1)$. It has already been shown that the higher i.r.'s, the D_p's, with p an integer, can be based on rather more complicated functions of x, y, and z (the solutions of Laplace's equation that are homogeneous in r), so they too can be used to generate operator i.r.'s, by using symmetrized forms. (While this is clearly possible, doing so becomes increasingly complicated as higher powers of r enter. Even a simple-looking expression, such as r^4, for example, contains x^2y^2. Its symmetrized replacement is quite a lengthy expression.) The similarity between the commutation properties of **L** and **J** means that the component of **L** can be replaced by the components of **J**, as convenient. The operator then acts in a space determined by **J**, so if J equals an integer plus one-half it gives an i.r. that acts in an orbital plus spin space. Nevertheless, the i.r.'s that are generated are still those of D_i where i is an integer.

§ 2.10 Spherically Symmetric Hamiltonians

With the concept of rotational symmetry it becomes possible to illustrate a few more uses of symmetry arguments in quantum mechanics. A common starting point is to use a quantum analogue of a classical Hamiltonian. This is usually not sufficient, for as experience has been gained it has become clear that there are many interactions in the quantum Hamiltonian of even the simplest conceptual classical systems which are not immediately apparent, the interactions arising because there is a magnetic moment associated with the electron spin being one example. Some way is needed of deciding what these additional interactions are and how to describe them, the most important test being, in the end, agreement with experiment. But more often than not it is a long path from the Hamiltonian to the interpretation of experimental results, so the usual procedure is to make a choice that seems sensible and see what happens. Any general principles that can be used as a guide are therefore welcome.

One commonly adopted principle is that if the Hamiltonian is to describe the properties of an assembly of electrons and nuclei then, for identical particles, it should be invariant under any permutation of their numbering. Another is that in free space the Hamiltonian should be invariant under any rotation of axes, for this should ensure that there are no preferred directions in space. So far, the free space Hamiltonians which have been discussed have included terms to describe the kinetic energies of the electrons, their Coulomb interactions with a single nucleus, and their mutual Coulomb interactions, with spin-orbit interactions being mentioned in passing. It is now of interest to consider the form that the latter should take.

The commutation relations for the components of the spin angular momentum operators of an electron are identical with those of the components of orbital angular momentum, with one important difference, that whereas the magnitude of the orbital angular momentum is allowed to have a variety of values, the magnitude of the spin angular momentum is fixed. So there are additional relations, such as $s_x s_y = i s_z$ and the similar relations obtained by cyclically interchanging x, y, and z. They show that any product of the spin components of a given electron can be reduced either to a constant or to an expression linear in s_x, s_y, and s_z. The scope for using the single-electron spin operators as bases for irreducible representations is therefore limited to bases for D_0 and D_1. The former is a constant and the latter is the set $\{s_x, s_y, s_z\}$. So any interaction in the Hamiltonian that contains the electron spin operators and is an invariant under rotational symmetry must take the form

$$u_x s_x + u_y s_y + u_z s_z,$$

where $\{u_x, u_y, u_z\}$ will form a basis for a D_1 representation. That this is necessary follows from the decomposition of $D_n \times D_m$ into irreducible representations, for it only contains D_0, the invariant representation, if $n = m$. If the spin-orbit interaction in free space is to be invariant under all rotations of axes it must therefore take the form $\mathbf{u} \cdot \mathbf{s}$, where, for a particular electron, $\mathbf{u}$ is a vector expression in orbital variables. For a many-electron system there is an extra consideration, for $\mathbf{u}$ and $\mathbf{s}$ do not have to refer to the same electron. So $\mathbf{u}$ can be a vector constructed from either the spin or the orbital variables of another electron. (There will also have to be a summation over the indices that number the electrons so that they are not distinguished.) This reasoning does not determine the form of $\mathbf{u}$, though as it transforms in the same way as $\mathbf{l}$ the spin–own-orbit interaction is often chosen to have the form $\zeta \mathbf{l}_i \cdot \mathbf{s}_i$, for the ith electron, with ζ being a parameter to be determined by experiment. The above reasoning suggests that there may also be spin-spin interactions, perhaps of the form $\mathbf{s}_i \cdot \mathbf{s}_j$ and orbit-orbit interactions of the form $\mathbf{l}_i \cdot \mathbf{l}_j$. The forms are not easily determined, for although the interaction between two magnetic dipoles is rotationally invariant, it does not take the form of a cosine interaction, $\mathbf{m}_i \cdot \mathbf{m}_j$, between two magnetic dipoles. Ignoring invariant factors, it has the form

$$[3\mathbf{m}_i \cdot \mathbf{m}_j - (\mathbf{m}_i \cdot \mathbf{r}_{ij})(\mathbf{m}_j \cdot \mathbf{r}_{ij})],$$

where $\mathbf{r}_{ij}$ is a unit vector. (With l_i and l_j able to take any integer values, many more invariants can be formed.) However, in the ions that will be of interest the spin–own-orbit interaction appears to dominate, so it is usual to neglect any others.

For a system containing just one electron the inclusion of the degree of freedom associated with electron spin doubles the degeneracy associated with each orbital state. The product states transform as $D_l \times D_{1/2}$, a reducible representation giving $D_{l+1/2} + D_{l-1/2}$. Without the spin-orbit coupling the degeneracy is not resolved, but with it a splitting into two levels, one with $j = l + 1/2$, with states transforming as $D_{l+1/2}$, and another with $j = l - 1/2$, with states transforming as $D_{l-1/2}$, is to be expected. This is observed, as are the predicted degeneracies. (It is typical of a group theoretical argument that neither the magnitude nor the sign of the splitting emerges without a knowledge of ζ, the spin-orbit coupling parameter.)

For a many-electron system which obeys Russell-Saunders coupling the magnitudes of the electron kinetic and Coulomb energies are such that in a first approximation the spin-orbit interactions can be ignored. Group theory then predicts that each energy level can be classified by an $\mathbf{L}$ and an $\mathbf{S}$ and that the states in it will transform reducibly, as

$$D_L \times D_S = D_{L+S} + D_{L+S-1} + \cdots + D_{|L-S|}.$$

With the spin-orbit interaction included the degeneracy is likely to be partially removed. There is then the further prediction that each level will be associated with one of the J values, which run in integer steps from $L + S$ to $|L - S|$ and that each J level will have a $(2J + 1)$-fold degeneracy. Since the L and S of the ground term of such an ion can usually be predicted from a knowledge of the eigenvalues and eigenfunctions of hydrogen, these predictions show the power of group theory. Indeed, it is not unreasonable to claim that the general pattern of the low-lying energy levels and the associated degeneracies of a many-electron ion that is of Russell-Saunders type can be obtained without needing to solve any quantum mechanical problem other than that of hydrogen.

A useful check is often obtained by applying a magnetic field, for this removes the spherical symmetry, leaving only axial symmetry. The irreducible representations of the new symmetry group are correlated with the states that are eigenstates of J_z, where Oz is taken in the direction of the field. The only symmetry operations left are arbitrary rotations about Oz, which change an eigenstate such as $|M_J\rangle$ into $\exp(iM_J\alpha)$, where α is the angle of rotation. All the irreducible representations are therefore one-dimensional and so it can be expected that all degeneracies will be lifted. Determining the degeneracies in zero field then becomes a matter of counting energy levels in the presence of a magnetic field and seeing how they collapse as the field is reduced to zero.

§ 2.11 The Wigner-Eckart Theorem

Much of the analysis given in this chapter has been leading up to the Wigner-Eckart theorem, which is probably the most widely used theorem in the application of group theory in theoretical physics. It is not necessary to give its proof here [it can be found, for example, in Judd (1963) and Thompson (1994)], nor will it be given as a formal statement, for it is a statement about relations between i.r.'s, and the application to physics requires some further steps. It is therefore easier to describe how it is used, by examples, of which there will be many in due course.

Its prime use is in the partial evaluation of matrix elements, where the inclusion of "partial" is deliberate because symmetry arguments cannot usually be used to evaluate quantities of interest, except in special cases when they happen to be zero. In these cases the theorem can be regarded as an extension of a well-known technique in integration, whereby a definite integral can sometimes be shown to be zero by a change of variable, such as the replacement of x by $-x$, which results in the integral being equal to its negative and so to zero. The change of variable can be regarded as the exploitation of a symmetry property of the integral. A related technique uses a suitable change of variables to turn one integral into another. The Wigner-Eckart theorem can be regarded as a sophisticated version of both of these techniques, for it provides a procedure for relating matrix elements, rather than evaluating them specifically.

A matrix element in quantum mechanics, such as $\langle a_i | v_j | b_k \rangle$, becomes an integral in wave mechanics and the Wigner-Eckart theorem can be used to examine how it changes if the bra vector $\langle a_i |$, the operator v_j, and the ket vector $|b_k\rangle$ all transform as elements in bases of i.r.'s of the same symmetry group. Suppose, for example, that $\langle a_i |$ belongs to a basis that transforms as Γ_i, that v_j belongs to a Γ_j basis, and that $|b_k\rangle$ transforms as Γ_k. Then one result of the theorem is that the matrix element vanishes unless the product

$$\Gamma_i \times \Gamma_j \times \Gamma_k,$$

when expressed in i.r.'s, contains A_1, the unit representation.

It is often incorrectly stated that the element vanishes unless $\Gamma_i \times \Gamma_j$ contains Γ_k. For any element B of a group its character, $\chi(B)$, in an outer product of two irreducible representations, is the product of its characters in each i.r. That is,

$$\chi(\Gamma_i \times \Gamma_j) = \chi(\Gamma_i) \times \chi(\Gamma_j)$$

for each B. The correct statement is that the product, when decomposed into i.r.'s, should contain the representation adjoint to Γ_k if the matrix element is not to vanish. The distinction between an i.r. and its adjoint is therefore important in using the theorem. In some groups the characters of one i.r. are the complex conjugates of the characters of another. The two i.r.'s are then said to be adjoints of one another. In many cases the distinction is unnecessary, for each i.r. is its own adjoint. It can, though, be disastrous if the theorem is used incorrectly.

The proof uses the property of a matrix element that it is a scalar and so is invariant under all the symmetry operations of the relevant group. The overall product $\Gamma_i \times \Gamma_j \times \Gamma_k$, when expressed as a sum of i.r.'s, either contains A_1, which is common to all groups (the 1×1 unit matrix), or it does not. Suppose, for example, that it contains a basis element of an i.r. which differs from A_1. Then it might appear that group theory is predicting that a sum of integrals changes its value as some specific rotation of axes is made, which cannot be correct. Since this argument can be applied to every i.r. in the reduced product, except A_1, the only escape is to accept that the integrals sum to zero if A_1 is missing. If it is present, group theory alone does not determine the value of the matrix element. However, it can then be used to show that a lot of other matrix elements have values that are numerically related to the present one, for the group operators can be used to turn the bra, the ket, and the operator between them into a variety of others. This provides a very useful technique.

Consider the matrix element $\langle a_i | v_j | b_k \rangle$, given above, and suppose that there is another operator, u_j, which transforms in exactly the same way as v_j and which is such that all its matrix elements, $\langle a_i | u_j | b_k \rangle$, when a_i and b_k run over their complete basis sets, can be determined. Then the theorem shows that the matrix elements of v_j are proportional to those of u_j, except in one special case, that in which all the $\langle a_i | u_j | b_k \rangle$ elements vanish.

An example is furnished by supposing that the $\langle a_i |$ and the $| b_k \rangle$ are states in a term of a two-electron system, so they all have the same L and S. The $\langle a_i |$ can then be written as $\langle L, M_i |$ and the $| b_k \rangle$ as $| L, M_k \rangle$. Suppose also that v_j is the Coulomb interaction, e^2 / r_{12}, between the two electrons. The bras and kets belong to i.r.'s that transform as D_L and $D_L \times D_L$ contains D_0, the invariant representation. The Coulomb interaction is rotationally invariant, D_0, so the triple product certainly contains D_0. Another invariant operator is the unit operator and for the same bras and kets its matrix elements are easily found. Indeed, they all vanish unless $M_i = M_j = M$, and then they are all equal to unity. It is therefore easy to compare the matrix elements of e^2 / r_{12} with those of the unit operator. When M_i is different from M_k the matrix element of the unit operator is zero and so, therefore, is the corresponding element of e^2 / r_{12}. When M_i equals M_k the unit operator has the same matrix element for all choices of M, so the corresponding elements of e^2 / r_{12} are all equal. The reasoning depends on the observation that under the operations of the group any matrix element involving v between two states transforms in exactly the same way as would be found with u replacing v. It follows that all the matrix elements of $\langle a_i | v_j | b_k \rangle$, within the manifolds determined by the basis sets, can be written in the form

$$\langle a_i | v_j | b_k \rangle = \langle a || v || b \rangle \langle a_i | u_j | b_k \rangle,$$

where $\langle a || v || b \rangle$ is a factor common to all the elements of v. It is known as a "reduced matrix element." A consequence of the theorem is that it tends to focus interest on reduced matrix elements. (The example, which has used a two-electron model, immediately extends to systems with any number of electrons. If the Coulomb interaction is replaced by a spin-independent Hamiltonian, which is also rotationally invariant, a simple extension shows that the Hamiltonian commutes with every element of $\mathbf{L}$. If the Hamiltonian contains spin operators the commutation is with the components of $\mathbf{J}$.)

Another example is to consider a matrix element $\langle a|V|b\rangle$, where $\langle a|$ belongs to a basis set for the T_1 i.r. of the cubic group, V to a basis set for A_2, and $|b\rangle$ to a basis set for E. From the character table $T_1 \times A_2 = T_2$, so $T_1 \times A_2 \times E = T_1 + T_2$. The unit representation does not appear, so it follows that not only does the matrix element vanish but that this will be true for all choices of $\langle a_i|$, V_j, and $|b_k\rangle$ that are symmetry related to $\langle a|$, V, and $|b\rangle$.

The above conclusion can be checked very simply. In considering transformations, $\langle a|$ can be identified with z, V with xyz, and $|b\rangle$ with $(x^2 - y^2)$. The matrix element therefore transforms in exactly the same way as $xyz^2(x^2 - y^2)$. The rotation that takes $x \to y$, $y \to -x$, and $z \to z$ reverses its sign, which shows that the matrix element is equal to its negative and so must be zero. The transformation $x \to y$, $y \to z$, $z \to x$ produces another matrix element, which will also be zero, as will be all other matrix elements that can be obtained by using symmetry operations. This reasoning requires no knowledge of group theory but simply the ability to spot what happens to a matrix element under certain substitutions. It can be used to illustrate a well-known comment "that there is nothing which group theory does which cannot be done without it." Nevertheless, the understanding provided by group theory of the role played by i.r.'s in the Wigner-Eckart theorem, which is yet to be fully demonstrated, illustrates that the time and effort involved in becoming familiar with certain aspects of group theory are a worthwhile investment.

All the examples so far have involved groups for which all the i.r.'s are self-adjoint. An example having i.r.'s that are complex is the group obtained by distorting a cube by stretching it along a body diagonal, so that the only remaining symmetries are rotations through multiples of $2\pi/3$ about this diagonal. Before the distortion one irreducible representation of the cubic group, T_1, has $\{x, y, z\}$ as basis. After the distortion the linear combination $(x + y + z)$ is clearly an invariant, so it forms a basis for an A_1 i.r. of the new symmetry group. There remain two further possible combinations of x, y, and z, which give either a two-dimensional i.r. or two one-dimensional i.r.'s. By inspection the combination $x + \omega y + \omega^2 z$, where $\omega = \exp(2\pi i/3)$, is the basis element for a one-dimensional representation, for the rotation $x \to y$, $y \to z$, $z \to x$ takes it into $\omega^2(x + \omega y + \omega^2 z)$, a multiple of itself. The character for this group element is therefore ω^2. The combination $x + \omega^2 y + \omega z$ is the basis for another i.r., and its character for the same group element is ω. One i.r. is therefore the adjoint of the other. (These combinations can be used to demonstrate the importance of the adjoint in the Wigner-Eckart theorem. If $|a\rangle$ is such that the above rotation multiplies it by ω, then $\langle a|$ is multiplied by ω^2. The product

$1|a\rangle$ is also multiplied by ω and since $\langle a|1|a\rangle$ is known to be nonzero it follows that $\langle a|$ must indeed be multiplied by ω^2, which is the character of the i.r. that is the adjoint of the i.r. for $|a\rangle$.)

§ 2.12 The Permutation Group

It has been stressed that in any many-electron Hamiltonian the electrons should not be distinguished. This is achieved by labeling them, say from 1 to N, and then adding the requirement that the Hamiltonian should be invariant under any rearrangement of the labeling. The symmetry thereby imposed can then be studied by using the i.r.'s of permutation groups, a technique that has been used extensively in the theory of many-electron atoms and ions (see Judd, 1963). However, in this book the properties of permutation groups, as such, will not be needed, for there is another way of incorporating the antisymmetry requirement, which is to use the method of second quantization. A description of this technique will be given in Chapter 6.

§ 2.13 Kramers' Degeneracy and Time Reversal

There is an entirely different type of symmetry associated with the time-dependent Schrödinger equation, which in wave mechanics is

$$i\hbar\frac{\partial\psi}{\partial t} = H\psi. \tag{2.17}$$

In the absence of an external magnetic field H is a real differential operator. If the sign of t is reversed and each side is replaced by its complex conjugate the result is the equation

$$i\hbar\frac{\partial\psi^*}{\partial t} = H\psi^*, \tag{2.18}$$

which is the same as eqn. (2.17) except that ψ has been replaced by ψ^*. To see the significance of the operation of reversing t and taking complex conjugates it is instructive to begin with a single-electron Hamiltonian and consider an eigenfunction of the Hamiltonian, taken in a form which explicitly includes its time dependence:

$$\psi = \exp\left(\frac{-iEt}{\hbar}\right)[f|+\rangle + g|-\rangle], \tag{2.19}$$

where f and g are functions of x, y, and z. The exponential factor is unaltered on reversing t and taking the complex conjugate, an operation which replaces f and g, respectively, by f^* and g^*. There remains the question of what happens to the $|+\rangle$ and $|-\rangle$ kets. As they are associated with components of spin angular momentum it might be expected that the operation will simply reverse the direction of the spin rotation, except that because spin has the curious 4π rotational invariance property, it might be as well to anticipate that there may be some associated phase changes. This point can be examined by using the properties of the s_x, s_y, and s_z operators with the assumption that they behave in exactly the same way as orbital angular momentum operators. That is, they simply reverse, so preserving the commutation rule $\mathbf{s} \wedge \mathbf{s} = i\mathbf{s}$ and the special property of spins of $1/2$ that $s_x s_y = i s_z$, etc. Suppose then that $|+\rangle$ goes into $\exp(i\alpha)|-\rangle$ and $|-\rangle$ goes into $\exp(i\beta)|+\rangle$. Under the operation of time reversal and complex conjugation $s_+ \to -s_-$ and $s_- \to -s_+$. Now $s_-|+\rangle = |-\rangle$ and $s_+|-\rangle = |+\rangle$, so it follows that $-\exp(i\alpha) = \exp(i\beta)$. The choice $\alpha = 0$ makes $\beta = \pi$, so establishing that there are phase changes associated with the operation. In particular, the state obtained by the transformation of (2.19) is the orthogonal state:

$$\psi^* = \exp\left(\frac{-iEt}{\hbar}\right) \exp(i\alpha)[f^*|-\rangle - g^*|+\rangle]. \tag{2.20}$$

It follows that each state with energy E has an associated orthogonal state of the same energy, so the energy level must have an even degeneracy. The pair of states are known as Kramers' conjugate states. With three electrons the wave function might contain a spatial and a spin part:

$$f|+-+\rangle + g|-+-\rangle \tag{2.21}$$

Under the operation it will turn into

$$f^*|-+-\rangle - g^*|+-+\rangle, \tag{2.22}$$

which is orthogonal to (2.21). Since all parts of a three-electron wave function can be paired in a similar way with parts of another wave function, it follows, again, that each energy level has an even degeneracy. The reasoning can be extended to any system with an odd number of electrons, but not to one with an even number. Thus Kramers' theorem only applies to a system with an odd number of electrons when no external magnetic field is present. It then states that all energy levels will have an even degeneracy. Even so, it has to be used carefully, for if the system also has an odd number of nuclei, each of which has spin $1/2$, it fails.

§ 2.14 References and Further Reading

There are many books on group theory, so a newcomer to the subject should probably choose the ones which seem easiest to understand. A source that covers most of the same ground as the present chapter is Chapter 7 of *Mathematical Techniques and Physical Applications*, by J. Killingbeck and G.H.A. Cole (1971) (Academic Press: New York). Other books on group theory are:

Cotton, F. A. 1990. *Chemical Applications of Group Theory* (Wiley: New York). This book contains the i.r.'s of all the point groups that are likely to be of interest, along with definitions of the symbols used to denote the various group operations).

Cracknell, A.P. 1968. *Applied Group Theory* (Pergamon: London).

Judd, B.R. 1963. *Operator Techniques in Atomic Spectroscopy* (McGraw-Hill: New York). This book contains the character tables of some of the point groups and a good deal of information about permutation groups.

Heine, V. 1960. *Group Theory in Quantum Mechanics* (Pergamon: London).

Thompson, W.J. 1994. *Angular Momentum* (John Wiley and Sons: New York).

Tinkham, M. 1964. *Group Theory and Quantum Mechanics* (McGraw-Hill: New York).

3

Perturbation Theory

§ 3.1 General Considerations

Most problems in quantum mechanics cannot be solved exactly, so recourse has to be made to approximation methods. There is no single method that is universally satisfactory, and over the years a variety of different procedures have been introduced and lumped together under the title of perturbation theory. In this chapter an outline will be given of some of the methods which are particularly relevant to magnetic problems and which will be used subsequently. First, though, there are several general remarks that can be made. Two types of Hamiltonian are in use. The first does not contain time explicitly, while the second, which is frequently used in dealing with excitations, is really an approximate Hamiltonian in which the radiation field has been taken in a classical form and with a specific time dependence. In this chapter it will be assumed that the Hamiltonian is of the first type and therefore is independent of time. It is used in the Schrödinger equation, which is a time-dependent equation of the form

$$i\hbar \frac{\partial}{\partial t} |t\rangle = H|t\rangle. \tag{3.1}$$

Being first order in t implies that its solutions are uniquely determined by the choice made for $|t\rangle$ at some specific time, t_0. It has particular solutions of the form

$$|E_n, t\rangle = \exp\left(\frac{-i E_n t}{\hbar}\right) |E_n\rangle, \tag{3.2}$$

where $|E_n\rangle$ is an eigenfunction of H with eigenvalue E_n. The general solution is then

$$|t\rangle = \sum_{n=0}^{\infty} A_n \exp\left(\frac{-i E_n t}{\hbar}\right) |E_n\rangle, \tag{3.3}$$

where the A_n are constants to be determined by the initial choice $|t_0\rangle$. Its form focuses interest on the eigenfunctions and eigenvalues of H. However, these cannot usually be found exactly. Nor are they all of equal interest, for only those of the lowest-lying energy levels are usually needed.

In most cases the Hamiltonian is separated into two parts, the unperturbed Hamiltonian H_0 and the perturbation V, equal to $H - H_0$. The latter is often thought of as $\lambda(H - H_0)$, so that it can be switched on by letting λ increase from 0 to 1. For any finite value of λ the expression $H_0 + \lambda(H - H_0)$ is often called the perturbed Hamiltonian. However, while it is easy to suggest that the Hamiltonian should be separated in this way it is not always easy to decide on the best division. Also, it is often assumed that some of the terms in H are smaller than others, so that they can be dropped in a first approximation. This does not necessarily mean that what is left is the unperturbed Hamiltonian, for it may still be too complicated to treat exactly. Having dropped the smaller terms, a perturbation approximation is then needed to deal with what is left, and this is usually an expansion in powers of λ, the assumption being that successive powers will be of decreasing magnitude when λ is set to unity. There is then a risk that some of the terms that have come from the expansion will be smaller than some of those originally dropped. Such considerations complicate perturbation theory, so since a typical Hamiltonian has a large number of eigenvalues and eigenstates, of which only a few are usually of interest, it is sensible to concentrate on these. Then it pays to use a formalism that can readily be extended to take account of interactions that have previously been dropped: the relative sizes of interactions are often determined a posteriori, by comparison with experimental results. Finally, all perturbation treatments have one feature in common, that any unperturbed Hamiltonian needs to be tractable, a requirement which severely limits the choice because there are not many that are.

§ 3.2 Nondegenerate Levels

The simplest case occurs when the energy levels of interest in the unperturbed Hamiltonian are nondegenerate. In this case there are infinite expansions in powers of λ for the eigenvalues and eigenfunctions of the perturbed Hamiltonian. Usually only the first few terms are given and used. To second order the energy of the ith level is

$$E(i) = E_0(i) + \langle i|\lambda V|i\rangle - \sum_{n \neq i}^{\infty} \frac{\langle i|\lambda V|n\rangle\langle n|\lambda V|i\rangle}{E_0(n) - E_0(i)}, \tag{3.4}$$

where $E_0(n)$ and $|n\rangle$ are, respectively, the nth eigenvalue and eigenstate of the unperturbed Hamiltonian ($\lambda = 0$) and the perturbation is λV. The various terms in the expansion are referred to as zeroth order (no dependence on λ), first order (linear in λ), and so on. There is a similar expansion for the perturbed eigenstate, which to first order in λ is

$$|I\rangle = |i\rangle - \sum_{n \neq i}^{\infty} \frac{|n\rangle\langle n|\lambda V|i\rangle}{E_0(n) - E_0(i)} \tag{3.5}$$

where $|I\rangle$ denotes the eigenstate of the perturbed Hamiltonian that is derived from the state $|i\rangle$ of the unperturbed Hamiltonian. In both expressions the right-hand side involves only the properties of the unperturbed Hamiltonian and the perturbation. There is a 1:1 relation between the energies (and states) of the unperturbed and perturbed Hamiltonians. As the perturbation is switched on, each level and state varies in a continuous way. It may be noticed that, for the lowest energy level, all the denominators in the λ^2 term are positive and the numerators are square moduli. Thus the second-order contribution to the lowest energy is never positive. A test of whether or not a perturbation is small is to compare the numerators with the denominators in the terms in the expansions.

Another feature common to both expressions is the occurrence of $|n\rangle\langle n|$. It is often replaced by an operator P_n, called a projection operator. Such operators have a number of useful properties. For example, P_n applied to a state $|m\rangle$, which is orthogonal to $|n\rangle$, gives zero, for $|n\rangle\langle n|m\rangle = 0$. When applied to $|n\rangle$ it simply reproduces $|n\rangle$. Using projection operators, the second-order perturbed energy can be written as

$$E(i) = E_0 + \langle i|V|i\rangle - \sum_{n \neq i} \frac{\langle i|V P_n V|i\rangle}{E_0(n) - E_0(i)}, \tag{3.6}$$

with a similar expression for the first-order perturbed eigenstate. The formulae can now be extended to cases where some or all of the levels, other than the one of interest, are degenerate, by replacing the projection operator for a single state by the projection operator of a set of degenerate states, which is the sum of the projection operators of all the states in that set. The projection operator of such a set is invariant under a unitary transformation of the states and so is a unique operator. Such a set is often called a "manifold of states." In many cases, such a degenerate set only occurs because of a symmetry in the unperturbed Hamiltonian and the states in the set form a basis for an irreducible representation of its symmetry group. The projection operator, an invariant, can be regarded as a basis for an A_1 i.r.

§ 3.3 Degenerate Levels

The perturbation expansion becomes more complicated if the unperturbed level of interest is itself degenerate, and the first step is to find the "correct states" for a zeroth-order approximation. To see what this means, it is useful to suppose that, for some finite value of λ, all the degeneracies have been lifted and the eigenvalues and eigenstates of the perturbed system have been found. All the perturbed states will be mutually orthogonal. This will continue to be so as λ is allowed to tend to zero. However, at zero, the states will be degenerate and so they can be arranged into degenerate orthonormal sets in an infinite number of ways. The limiting procedure would have shown which particular set should have been chosen, starting from $\lambda = 0$, except that there would then have been no way of knowing how to choose them. The chances are that if an arbitrary choice had been made it would not have been the correct one. So it is necessary to have a way of choosing the correct zeroth-order states.

The problem is solved in principle by setting up the matrix of the perturbation within the degenerate manifold of unperturbed states, using any convenient orthonormal set of them. Then the set of correct zeroth-order states is that which results in the matrix of λV becoming diagonal, and the procedure for finding them reduces to diagonalizing a matrix. With the aid of computers there is no problem in diagonalizing quite large matrices when the elements are given numerically, but when they are given as algebraic expressions, as often happens in perturbation theory, the matter of choosing the correct zeroth-order states can present a major problem. However, if they are found there is generally little difficulty in going on to the next order in perturbation theory, except in a relatively small number of cases when the perturbation consists of two parts, V_1 and V_2, with V_1 giving a small contribution in first order and a part which is larger in second order than that given by V_2 in first order. It may then be incorrect to choose the zeroth-order states using the matrix of V_2. Examples where this happens can be found in crystal field theory, and it is one of the reasons why all perturbations should be treated, if possible, on an equal footing.

§ 3.4 Effective Hamiltonians

There is an alternative way of dealing with a degenerate manifold in the unperturbed Hamiltonian, which is now used in a good deal of theoretical work in magnetism and which helps with the problem mentioned above. Yet it is seldom described in introductory accounts of perturbation theory. This section will therefore be devoted to explaining its general features, for it will be used in subsequent chapters.

When a degeneracy is completely lifted, the number of perturbed states is exactly the same as the number of orthonormal states in the unperturbed degenerate manifold. So if the interest is in the splittings that have come from this manifold (it is usually the manifold of lowest energy in the unperturbed system), it can be expected that the set of energy levels that have arisen from the perturbation will not overlap any energy levels that have come from other unperturbed manifolds. There is then the prospect of being able to define an "effective Hamiltonian," an operator which acts only within the lowest unperturbed manifold and yet which has eigenvalues that coincide with those expected to come from it when the perturbation is switched on. The eigenstates of the effective Hamiltonian will be linear combinations of the states in this unperturbed manifold, so they cannot be eigenstates of the perturbed Hamiltonian, which will contain admixtures of states that are outside this manifold.

To describe this manifold more succinctly, it is useful to extend the concept of a projection operator from that of a single state to that of a manifold of degenerate states. If the manifold contains r orthonormal states there are many different ways in which they can be defined. If a specific choice, $|1\rangle$, $|2\rangle$, ..., $|r\rangle$, is made, the projection operator for the manifold can then be defined as the sum of the projection operators of the r states. It is then readily seen that it is invariant to the choice and so it expresses a rather special property of the degenerate manifold. It is usual to use a notation for it, P, with some suitable subscript that is similar to that of the projection operator of a single state. This seldom leads to confusion because, in the perturbation expansions, if states in a degenerate manifold appear there are invariably summations over all states in that manifold such that the result of the summation is invariant to the choice of the r states. The summation can then be replaced by the introduction of the projection operator for the manifold, a step that tidies up the perturbation expression. That it will happen can be anticipated because there is usually no reason to suppose that any particular combination of states in a degenerate manifold will emerge from a formal expression of perturbation theory.

The problem of how to invent an effective Hamiltonian which acts within a specific manifold of states, which can be denoted by P_0, the projection operator for the manifold, and which reproduces the pattern of eigenvalues that emerges from it under a perturbation was investigated by Bloch (1958). He produced two different treatments, the first of which is described at length by Messiah (1962). The second appears to be less well known, and yet it seems to be of wider generality, for its derivation makes less restrictive assumptions. This will not be given here, but only the result. The unperturbed Hamiltonian is taken to have the form

$$H_0 = \sum_n E_n P_n, \tag{3.7}$$

where P_n is the projection operator of a degenerate manifold of states with energy E_n, and n takes integer values 0, 1, 2, etc. The perturbation V is then the difference $(H - H_0)$ between the Hamiltonian of interest and the unperturbed Hamiltonian. The effective Hamiltonian, which gives the eigenvalues into which P_0 is split, is an infinite expansion in powers of V, the first few terms of which are

$$H_e = P_0 \left[E_0 + V - \sum_{n \neq 0} \frac{V P_n V}{(E_n - E_0)} + \ldots \right] P_0. \tag{3.8}$$

On the right- and left-hand sides the expansion is bordered by a P_0 projection operator, which ensures that the effective Hamiltonian can be regarded as an operator in the reduced Hilbert space defined by P_0. When just a few terms are given, the expression looks similar to that of standard perturbation theory. A full expansion reveals differences, one being that the only energy denominators which appear are those between energies of excited levels and E_0. There are no denominators that involve differences between intermediate energy levels, E_n and E_m, with n and m both different from zero.

The method still does not give the energy levels directly. Instead, it leaves a problem similar to that of diagonalizing the zeroth-order matrix of standard degenerate perturbation theory, with the difference that having done so the energies are given, in principle, to all orders in perturbation theory. (In practice the expansion is invariably terminated at some finite order.)

A nice feature of the technique is that it produces an operator which acts in a subspace of the Hilbert space of the actual Hamiltonian. If this space has dimensions N, then N can be set equal to $(2J + 1)$, where J is either an integer or an integer plus one-half. The effective Hamiltonian can then be expressed in the angular momentum variables of a moment of magnitude J. That this is possible can be seen by using Oz as the axis of quantization and considering the operators 1, J_z, J_z^2, J_z^3, etc. They are all diagonal and there are sufficient of them to make it possible to choose a linear combination that will match the diagonal elements of any $N \times N$ matrix. Similarly, the set J_+ and J_- times the different powers of J_z can be combined to match all the one-off diagonal elements, and so on, using higher powers of J_+ and J_-. Once having got the effective Hamiltonian into angular momentum form, there is no longer any need to use Oz as the axis of quantization, particularly if some more convenient axis exists.

In the examples that will be described the method always produces multiples of the unit operators of the various P_n families, which describe their displacements relative to the chosen E_n values. These are of no significance when the interest is in *the separations between the energy levels* into which a given P_n is split by the perturbation, as will usually be the case. Such constant terms will therefore be systematically dropped. So where, above, it is stated that the Bloch method gives an expression that determines the eigenvalues into which a given degenerate family is split, it will more often be used to give an expression in which each energy has had the same constant value subtracted from it. It will then give the energy level splitting pattern which would be observed if all the other energy levels had been removed.

§ 3.5 Creation Operators

There are many other perturbation techniques, a particularly important one being based on the idea of excitations from an assumed nondegenerate ground state of the actual Hamiltonian. A set of operators is sought, such that when any one of them is applied to the ground state it produces an excited state, an eigenstate of the Hamiltonian, at a known energy above that of the ground state. If two such operators are applied in succession, then there can be an expectation that a new eigenstate will be obtained at an energy above the ground state which is a sum of the excitation energies of the two operators, and so on. The perturbation technique is aimed at finding these operators, their commutation properties, and the excitation energies that go with them. The approximations enter in various ways, the first being that such operators exist. Usually, the ones which are chosen do not have precisely the properties sought, so the assumption is then made that in some sense they are good enough. One of the ideas behind this procedure is that with an extended system it is reasonable to expect that it can be excited in two or more different places, and that such excitations will not interfere with one another. The operators and their associated excitation energies are used to define effective Hamiltonians, which are then used to predict low-temperature thermodynamic properties. As this involves statistical concepts, such as averages over distributions over energy levels and the concept of temperature, it frequently seems that the chosen operators are good enough to account for the experimental observations.

§ 3.6 Thermodynamic Functions

A closely related technique is to design a perturbation theory in which thermodynamic features are already incorporated, only now with the creation and annihilation operators being defined with respect to the unperturbed Hamiltonian. This is basically more satisfactory and somewhat like the effective Hamiltonian approach, for it concentrates attention on a selection of eigenvalues, those with energies that are not more than a few kT above the ground state. The Hamiltonian is split into an unperturbed Hamiltonian and a perturbation, and a perturbation expression is developed, in powers of the perturbation, to give a thermodynamic quantity, such as the partition function or the free energy.

The partition function, Z, will be chosen as an example. It is defined by

$$Z = \sum_n \exp(-\beta E_n) = \text{Trace} \left[\exp(-\beta H)\right], \qquad (3.9)$$

where $\beta = 1/kT$, T being the temperature, and the E_n are the eigenvalues associated with each eigenstate of the Hamiltonian. The summation is over all eigenstates, not just the eigenvalues. The next step is to define two resolvent operators, one for the actual Hamiltonian and one for the unperturbed Hamiltonian, H_0. They are defined in similar ways and each contains a complex variable, z. The definitions are

$$R(z) = (z - H)^{-1}, \quad R_0(z) = (z - H_0)^{-1}. \qquad (3.10)$$

Then

$$Z = \frac{1}{(2\pi i)} \oint \sum_n \exp(-\beta z) \frac{1}{(z - E_n)} dz$$

$$= \frac{1}{(2\pi i)} \oint \exp(-\beta z) \, \text{Trace} \, R(z) dz, \tag{3.11}$$

where the contour integral is taken round any closed curve that encircles all the poles of the integrands, which are at the values $z = E_n$. The relation

$$(z - H)R(z) = 1 = (z - H_0)R_0(z) \tag{3.12}$$

is then used to show that $R = R_0 V R + R_0$, an expression that can be iterated, by first replacing R in the first term on the right-hand side by $R_0 V R + R_0$ and then continuing the process indefinitely to give

$$R = R_0 + R_0 V R_0 + R_0 V R_0 V R_0 + \dots. \tag{3.13}$$

On substituting this expression into (3.11) the partition function becomes a contour integral of a sum of traces of products containing H_0 and V. Knowing the eigenstates and eigenvalues of H_0, each term in the expansion can, in principle, be determined, for the integrations reduce to sums of poles. In practice the terms in the expansion become successively more complicated to evaluate, and it may be better to begin with an expression for some other thermodynamic function, such as the free energy. Summations will again occur, but they may be expressible in terms of Feynman diagrams, which allow the evaluations of parts of successive terms to be carried to infinite order. [A detailed discussion of these techniques can be found in specialized texts such as that by Fetter and Waleka (1971).]

§ 3.7 A Simple Example

A good deal of experience about perturbation theory can be obtained by studying simple examples, and a convenient first choice is that of finding the low-lying energy levels of the free Ti^{3+} ion, which has nineteen electrons. In its ground configuration all but one of the electrons are in filled shells, so its properties are dominated by a single electron which is in a $3d$ state outside the filled shells. The ion has, of course, many excited configurations, each having a different set of occupation numbers for its hydrogenic-like one-electron states. The ground configuration has tenfold degeneracy, for there are five orbital states in the $l = 2$ of the $3d$ electron and a further factor of 2 comes from its spin of $1/2$. The orbital parts transform as D_2 and the spin as $D_{1/2}$, so as $D_2 \times D_{1/2} = D_{5/2} + D_{3/2}$ it can be expected that the tenfold degeneracy will be split into multiplets characterized by $J = 5/2$ and $3/2$ by any interaction that couples the orbital motion to the spin, such as the spin-orbit coupling. This is as far as can be gone using group theory alone. To determine which of the two J levels is lower and their separation, it is necessary to turn to perturbation theory.

The Bloch formalism will be used, so the first requirement is to choose the unperturbed Hamiltonian, the P_n projection operators, and their E_n coefficients [eqn. (3.7)]. Knowing the latter is not immediately necessary, however, if it is a first approximation to the splitting pattern of the ground term that is required, for the first term in the expansion, $P_0 H P_0$, does not involve the E_n's which have $n > 0$. For the projection operators the choice is limited; about all that can be done is to choose them to span the various configurations. P_0, the projection operator of the ground configuration, will therefore need to span ten states, which means that the first term in the perturbation expansion for this configuration will be a 10×10 matrix. (Each P_n is actually an operator in a much larger Hilbert space, that of all the eigenstates of Ti^{3+}. Each, however, takes a block diagonal form consisting almost entirely of zeros, except for the relatively small number of states that it spans, when it has unity on the diagonal. A similar block diagonal form is taken by any operator bordered to left and right by the same P_n. The matrix can therefore be regarded as representing an operator in the much smaller Hilbert space determined by the nonzero block.) Diagonalizing a general 10×10 matrix is usually troublesome; this example is an exception, for group theory has indicated that it should be possible to diagonalize it by choosing the states in P_0 so that they form bases for $D_{5/2}$ and $D_{3/2}$ i.r.'s. In fact, even this is not necessary, for one state in $J = 5/2$ and another in $J = 3/2$ will be enough. One of these is readily found, for the state which has $M_L = 2$ and $M_S = 1/2$ has $M_J = 5/2$ and so belongs to $D_{5/2}$. (The electrons in closed shells give orbital and spin states which transform as D_0. They play no role in the determination, in first order, of the splitting of the configuration.) To find a state in $J = 3/2$ the first step is to find the state in $J = 5/2$ which has $M_J = 3/2$. Its unnormalized form is obtained by applying J_-, which equals $L_- + S_-$, to $|5/2, 5/2\rangle$, where the first number gives J and the second M_J. The result is readily obtained when the state is expressed as $|2, 1/2\rangle$, where the first number now gives M_L and the second M_S. The result is the normalized form:

$$|J = 5/2, 3/2\rangle = \frac{1}{\sqrt{5}}[2|1, 1/2\rangle + |2, -1/2\rangle]. \tag{3.14}$$

The state in $J = 3/2$ with $M_J = 3/2$ is orthogonal to (3.14) and contains the same M_L, M_S states. Ignoring an ambiguity in phase, which is of no significance, it follows that

$$|3/2, 3/2\rangle = \frac{1}{\sqrt{5}}[|1, 1/2\rangle - 2|2, -1/2\rangle]. \tag{3.15}$$

So first-order approximations to the eigenvalues are given by the expectation values of the Hamiltonian taken over these two states. (To complete such a calculation it is necessary to know the radial parts of the one-electron wave functions. When such calculations are initially carried out these are seldom known, so the first-order splitting is an expression containing unknown quantities. The next step is then to compare the theoretical results with experimental observations to obtain approximate values for the unknown quantities with, later, an attempt to justify the observed values by some other method, such as a variational method, to obtain the "best" radial functions.)

A simplification is to focus on obtaining an expression for the splitting, rather than the energy of each multiplet. The Wigner-Eckart theorem can be used to replace the spin-orbit coupling, $\mathbf{u} \cdot \mathbf{s}$, summed over all electrons, by $\lambda \mathbf{L} \cdot \mathbf{S}$, for $\mathbf{u}$ and $\mathbf{s}$ provide bases for D_1 representations, one in the orbital and the other in the spin space. ($\mathbf{u}$ is replaced by a multiple of $\mathbf{L}$ and $\mathbf{s}$ by a multiple of $\mathbf{S}$, for each electron. With one electron outside a closed shell $L = 2$ and $S = 1/2$, the values of l and s of the electron in the $3d$ shell.) The relation

$$2\mathbf{L} \cdot \mathbf{S} = \mathbf{J} \cdot \mathbf{J} - \mathbf{L} \cdot \mathbf{L} - \mathbf{S} \cdot \mathbf{S} = J(J+1) - L(L+1) - S(S+1)$$

can then be used to show that, with λ positive, $J = 5/2$ is higher than $J = 3/2$ by $5\lambda/2$. Experiment confirms that λ is indeed positive, which is the same sign as that found for the $3d$ electron in hydrogen.

All orders in the perturbation expansion, beyond the first, involve choosing the P_n's and their associated E_n's. In the present example the obvious choice is to use the configurations to define the P_n's, which leaves the question of how to define the E_n's unanswered. In a general case an indication of how both might be obtained begins by noting that with sufficient insight it might be possible to choose the states in the P_n's so that they are actually eigenstates of the Hamiltonian, in which case the expectation values of the Hamiltonian taken over them would be eigenvalues. It can then be reasoned that if the states which are actually used in defining the P_n's are good enough approximations to the actual eigenstates, the expectation values of the Hamiltonian, taken over them, will be good approximations to the true eigenvalues. On second thoughts though, this procedure can be seen to have a drawback, for the essence of the Bloch perturbation method is that the P_n's describe degenerate sets of states in an unperturbed Hamiltonian, which is the reason why each state in P_n is given the same E_n. Taking expectation values of the Hamiltonian over states that are thought to approximate to eigenstates is therefore unlikely to result in each state in the same P_n having the same expectation value. The difficulty can easily be circumvented, by choosing the E_n that is associated with a specific P_n as equal to the mean of the expectation values of the Hamiltonian taken over the states chosen to define P_n. In the present example, with P_n chosen to span the states in a specific configuration, the associated E_n would then be the mean energy of each configuration of the Hamiltonian.

Such a procedure has several attractive features. The actual Hamiltonian, rather than some model Hamiltonian, is being used in the definition of the unperturbed Hamiltonian and the E_n's are likely to be widely spaced, so that the $(E_n - E_0)$ denominators in the perturbation expansion will be large. Each P_n will be invariant to the states used in its definition, which suggests that when choosing them it would be sensible, if at all possible, to arrange them into sets which provide bases for i.r.'s of the symmetry group of the Hamiltonian. This will then make each P_n, the unperturbed Hamiltonian and the perturbation, the difference between the Hamiltonian and the defined unperturbed Hamiltonian, invariant under the symmetry group of the Hamiltonian. (Such a procedure does not remove the need to choose "correct zeroth order states"; it simply postpones it.)

In most cases the Hamiltonians used are approximate Hamiltonians, so at this point it seems appropriate to emphasize that an unperturbed Hamiltonian defined by the above procedure is not the same, except by chance, as an approximate Hamiltonian. The latter will be regarded as obtained from the actual Hamiltonian by dropping interactions because they are thought to be, in some sense, small. The unperturbed Hamiltonian will be regarded as an auxiliary Hamiltonian which is of no physical interest. It is an operator that has been introduced solely as a mathematical convenience. It could just as well have been defined for an approximate Hamiltonian and indeed it usually is. Its only virtue, which itself is perhaps questionable from a physical point of view, is that it provides a way of introducing preconceived ideas about the wave functions, ideas that are inevitably incorrect, for going beyond first order in the Bloch perturbation formalism is a disguised way of correcting the choice.

Returning to Ti^{3+} the numerators in the second-order terms in the perturbation expansion can be regarded as describing a sum of "processes." Reading from right to left each process first uses a term in the Hamiltonian to connect a state in P_0 to a state in a P_n and then the same or another interaction connects the state in P_n to the initial, or some other state, in P_0. Since both the Hamiltonian and the unperturbed Hamiltonian have rotational symmetry, the perturbation will also have this symmetry. It will, therefore, only connect states in the same D_J irreducible representations and, within these, only states with the same M_J (by the Wigner-Eckart theorem). Since the same feature is present in all higher orders, it follows that there will be no further resolution of the $J = 5/2$ and $J = 3/2$ multiplets, though there will almost certainly be changes in their separation and in their states. This example is an illustration of the power of group theory, for it has easily produced a result that might otherwise only have emerged after an extensive investigation of an infinite expansion generated by perturbation theory. The whole procedure can be regarded as developing an operator that acts within P_0 and which is rotationally invariant. Since P_0 is itself spanned by $L = 2$ and $S = 1/2$ the invariant operators which are nonzero within it are severely restricted, to those that can be constructed using $\mathbf{L}$ and $\mathbf{S}$. With $\mathbf{S} = 1/2$ the only possibilities are constants and multiples of $\mathbf{L} \cdot \mathbf{S}$. (Either capital or lower-case letters can be used to denote angular momentum operators, for the only need is to span a Hilbert space of ten states. It would be possible to use a single operator, $J = 9/2$, for $2J + 1 = 10$. However, in carrying out the simplification of the various orders of perturbation theory it soon becomes obvious that the $L = 2$, $S = 1/2$ choice is a more natural one.) With either choice the bordering P_0's can be dropped, to leave, with the $L = 2$, $S = 1/2$ choice an "effective Hamiltonian"

$$H_{\text{eff}} = E_0' + a\mathbf{L} \cdot \mathbf{S},$$

which contains just two parameters, E_0' and a. They can be regarded as quantities either to be found by experiment or to be calculated using the perturbation expansion. Since $2\mathbf{L} \cdot \mathbf{S} = J(J + 1) - L(L + 1) - S(S + 1)$, the effective Hamiltonian can be further simplified so that the only variable it contains is J.

The application of an external magnetic field eliminates the rotational symmetry so it can be expected that all degeneracies will be resolved. For the ground manifold there is then an interest in whether or not energy levels coming from different J manifolds will cross. If they do a 10×10 matrix will need to be diagonalized, whereas if they do not then two matrices, one 4×4 and the other 6×6 will need to be diagonalized, which looks a simpler task. With the assumption that the field is too weak to produce crossings of states from adjacent J levels the projection operators can be chosen to be those appropriate to the various multiplets, the J states in each P_0. The state in the $J = 3/2$ ground multiplet which has $M_J = 3/2$ is given by (3.15) and the other three can be found from it, so the P_0 for this level is known. To first order in perturbation theory the splitting induced in it by the Zeeman part of the magnetic interaction, $\beta B(l_z + 2s_z)$, summed over all electrons, is found by diagonalizing this operator within P_0. Since the states in P_0 span a $D_{3/2}$ i.r. and the Zeeman operator transforms as D_1, it can be replaced by an equivalent operator which also transforms as D_1. The obvious choice is J_z, with a reduced matrix element which, conventionally, is denoted by Λ. So instead of giving the energy levels explicitly there is the alternative of stating that, to first order in perturbation theory, the Zeeman splitting of $J = 3/2$ is given by an operator, $\lambda\beta B J_z$, acting within $J = 3/2$, where Λ can be determined using a specific state in $J = 3/2$, the obvious choice being that given in eqn. (3.15). Using λJ_z on the left and $\lambda(L_z + 2S_z)$ on the right it follows that $\Lambda = 4/5$. The axes can then be eliminated by replacing $\Lambda\beta B J_z$ by $\Lambda\beta\mathbf{B} \cdot \mathbf{J}$. It is fairly simple to go on to second order and develop terms that are linear (from a cross product of Zeeman and spin-orbit interactions) and quadratic in $\mathbf{B}$. Doing so, and showing that the effective Hamiltonian will have rotational symmetry in the sense that the same splitting will be obtained whatever the direction of $\mathbf{B}$, is left to the reader as an exercise, as is finding the value of Λ for the $J = 5/2$ manifold.

§ 3.8 Many-Electron Ions

It usually happens, for the ions of interest in magnetism, that the lowest electronic configuration is more highly degenerate than that of Ti^{3+}, because there are more electrons in the $3d$-shell. The degeneracies are associated with the wider range of possible values for L and S. To examine this in more detail it is convenient to consider Cr^{3+}, which has three $3d$-electrons. Its ground configuration has a degeneracy of 120. They can be arranged into terms, manifolds with a variety of L and S values. From the published tables (see Condon and Shortley, 1967, for example) this configuration gives rise to the following terms: ^2P, D_2, F, G, H, ^4P, F, where the subscript 2 on D means that there are two such terms, and the superscript before each family of capital letters denotes the multiplicity, the value of $(2S + 1)$, for all the terms in that family. (P, D, F, G, and H denote $L = 1, 2, 3, 4$, and 5, respectively.) The various terms can be expected to have different energies and indeed, by calculating the expectation values of the Coulomb interactions in the Hamiltonian for selected states in the different terms, it is possible to show that ^4F is the term of lowest energy (a result which agrees with one of Hund's empirical rules). This term has $L = 3$ and $S = 3/2$ so it gives rise to multiplets with $J = 9/2, 7/2, 5/2$, and $3/2$, using

$$D_3 \times D_{3/2} = D_{9/2} + D_{7/2} + D_{5/2} + D_{3/2}.$$

When the spin-orbit coupling is omitted they will be degenerate, so, if they are used to define the lowest P_0, finding the first-order splitting due to the spin-orbit coupling, $\mathbf{u} \cdot \mathbf{s}$ summed over all electrons, requires the diagonalization of a 28×28 matrix. However, since all the orbital parts of the states in P_0 come from $L = 3$ and all the spin parts come from $S = 3/2$, the $\mathbf{u}$ for each electron can be replaced by a multiple of $\mathbf{L}$ and the $\mathbf{s}$ by a multiple of $\mathbf{S}$. The spin-orbit coupling within P_0 then becomes equivalent to $\lambda \mathbf{L} \cdot \mathbf{S}$, where λ is a reduced matrix element, known as the spin-orbit coupling parameter. Using the same formula as before: $\mathbf{J} \cdot \mathbf{J} = (\mathbf{L} + \mathbf{S}) \cdot (\mathbf{L} + \mathbf{S}) = \mathbf{L} \cdot \mathbf{L} + \mathbf{S} \cdot \mathbf{S} + 2\mathbf{L} \cdot \mathbf{S}$, shows that $\mathbf{L} \cdot \mathbf{S}$ within the ground term is equivalent to $(1/2)[J(J + 1) - 12 - 15/4]$, with J taking the values $9/2, 7/2, 5/2$ and $3/2$. The spin orbit coupling therefore splits the ground term into four multiplets with the energy differences between them being in the ratios $9 : 7 : 5$. (Such a numerical result can usually be tested fairly easily by experiment. In the history of spectroscopy, ratios have often been discovered by the experimentalists before the explanations have been forthcoming.)

A variety of rules about the energies of terms in a given configuration and their J splittings can be found in Condon and Shortley's book. On the whole, the results agree so well with the predictions from first-order perturbation theory that there is little need to go to higher order, particularly when group theory shows that no further splittings will occur on doing so.

A further prediction is that to any order in perturbation theory the effective Hamiltonian for the ground term of Cr^{3+} will consist of a constant, a term linear in $\mathbf{L} \cdot \mathbf{S}$, a term in $(\mathbf{L} \cdot \mathbf{S})^2$, and another in $(\mathbf{L} \cdot \mathbf{S})^3$, with no higher powers of $(\mathbf{L} \cdot \mathbf{S})$. This follows because with $S = 3/2$ any expression containing a component of $\mathbf{S}$ raised to a power higher than 3 can be reduced to one of lower power, which limits the range. A formal way of verifying the prediction is to write down the basis elements of the various D_n's, normalized in the same way as the Y_n^m's, using the components of $S = 3/2$. Then for each basis set another can be formed by replacing each component of $\mathbf{S}$ by a corresponding component of $\mathbf{L}$. Each element in the S basis can be denoted by $|S, M_S\rangle$ and the corresponding element in the L basis by $|L, M_L\rangle$ with $S = L$ and $M_S = M_L$. The product of $\langle L, -M_L|$ and $|S, M_L\rangle$ summed over all M_L values is an invariant, the D_0 component of $D_S \times D_L$ with $S = L$. There will be three such invariants, apart from a constant, because the components of $\mathbf{S}$ used once give a basis for D_1, used twice give a basis for D_2, and used three times give a basis for D_3, and this is all. It is possible, though not essential, to rearrange these invariants into expressions in powers of $\mathbf{L} \cdot \mathbf{L} = L(L+1)$, $\mathbf{S} \cdot \mathbf{S} = S(S+1)$ and $\mathbf{L} \cdot \mathbf{S}$ [see table 3 of Stevens (1992), where $\mathbf{T}$ is used in place of $\mathbf{L}$]. The parts in $\mathbf{L} \cdot \mathbf{S}$ can then be further simplified by expressing them in terms of J, L, and S.

The application of a magnetic field will remove all degeneracies. To include the Zeeman term it is necessary to add $\beta B(l_z + 2s_z)$, summed over all electrons, to the Hamiltonian. In first order of perturbation theory this is replaced, within a manifold of definite J, by another Λ times J_z, where, as for the Ti^{3+} case, Λ depends on J. Each multiplet is predicted to split into a pattern of equally spaced levels. Counting their number will establish the value of J, and a determination of Λ gives an additional check of the accuracy of the theory. Discrepancies between theory and experiment can obviously occur because the theory has only been given to first order, and even then it has used only the terms linear in the magnetic field in the Hamiltonian. Going on to higher orders in perturbation theory is certainly possible. In second order, the Zeeman terms will produce terms in the effective interaction that are linear in the magnetic field, from cross terms between the Zeeman interaction and the spin-orbit interaction, which will modify Λ, and terms quadratic in the magnetic field. It will then be necessary to include in first order the terms which are quadratic in B. While all this will complicate the form of the effective Hamiltonian it will nevertheless still be an invariant, for the splittings do not depend on the direction of $\mathbf{B}$. Thus the matter of choosing the correct zeroth order states does not arise. Any convenient choice can be made; the energy levels are simply the eigenvalues of the effective Hamiltonian.

The Ti^{3+} example can be used to illustrate another point, for titanium has two naturally occurring isotopes with different nuclear spins, 5/2 and 7/2. With either isotope the full Hamiltonian will contain interactions between the nuclear spin, $\mathbf{I}$, and the electrons. Such interactions are usually neglected in approximate Hamiltonians, in which case the nuclear spin operators, the components of $\mathbf{I}$, do not appear. Nor do they appear in the unperturbed Hamiltonian. Yet the dimensions of the Hilbert spaces for all the Hamiltonians, exact, approximate, and unperturbed, have been increased by a factor of $(2I + 1)$, where I differs for the two isotopes. The difference is revealed when the operators that couple the nuclear spin to the electron are included in the Hamiltonian. With $I = 0$ it is usually convenient to use angular momentum operators, such as $\mathbf{J}$ or $\mathbf{L}$ and $\mathbf{S}$ to span P_0, with the result that the effective Hamiltonian is an expression in these angular momentum operators. When I is nonzero it is usually convenient to include $\mathbf{I}$ in the set.

There is an interesting exception to the above conclusion, which can be illustrated for an isotope that has $I = 0$, as occurs in the most abundant naturally occurring isotope of chromium. In this case an approximate Hamiltonian is obtained by neglecting all the interactions that involve the electron spin. The P_0 of the unperturbed Hamiltonian is required to span a Hilbert space that has dimensions determined by the degeneracy of a term, which, because these two operators are related by the effect of the exclusion principle, is not independent of the total spin. So the effective Hamiltonian for a specific term may contain the components of $\mathbf{S}$ even though $\mathbf{S}$ does not appear in the generic Hamiltonian. (This feature is show in a particularly striking way when a Cr^{3+} ion is placed in an octahedral environment in a crystal. The effective Hamiltonian, then called the spin-Hamiltonian, only contains the components of $\mathbf{S}$, even though none of these occur in the approximate Hamiltonian.)

§ 3.9 References

Bloch, C. 1958. *Nucl. Phys.* 6:329.

Condon, E.U., and Shortley, G.H. 1967. *The Theory of Atomic Spectra* (University Press: Cambridge).

Fetter, A.L., and Walecka, J.D. 1971. *Quantum Theory of Many-Particle Systems* (McGraw-Hill: New York).

Messiah, A. 1962. *Quantum Mechanics* (North-Holland: Amsterdam), vol. II, p. 712 onwards.

Stevens, K.W.H. 1992. *Physica* B 172:1.

§ 3.10 Further Reading

Brandow, B. 1977. *Adv. Phys.* 26:651, §6.

4

Crystal Field Theory

§ 4.1 The Crystal Field Concept

Crystal field theory is widely used in explaining the magnetism of crystals, particularly in its spin-Hamiltonian form. It was introduced by Van Vleck in the early 1930s (see Van Vleck, 1944) because the ionic crystals of the iron group elements had been found to show magnetic susceptibilities which varied inversely as the temperature with magnitudes that could not be explained unless their orbital magnetism was neglected. There was, therefore, a need to demonstrate that on placing an ion in a crystal lattice the electronic motion would be so changed by the internal electric field set up by the neighboring ions, which, in the crystals then being studied, usually consisted of six O^{2-} ions arranged to form a distorted octahedron, that the orbital moment would disappear. A simple classical estimate indicated that the crystal field interaction, as the effect was described, would have an associated energy larger than that associated with the spin-orbit interactions and smaller than the Coulomb interactions. In any sequence of approximations it therefore seemed that the crystal field interaction should come after the interactions that produce the term structures of the free ions and before those that produced the multiplets. Also, as magnetic phenomena typically involve comparatively small values of kT, it was assumed that only the energy levels coming from the lowest term would be thermally populated, which implied that attention could be focused on a limited number of energy levels.

The general problem of how a perturbation of octahedral symmetry would split a D_L i.r. had already been studied, using group theory, by Bethe (1929), who showed that the degeneracies of spherical symmetry could be expected to be resolved as the symmetry was reduced. The theoretical need was to put this on a numerical basis and so obtain more detail about the resultant energy level patterns and wave functions.

An examination of the ground terms of the $3d^n$ configurations shows that the only L values which occur are 0, 2, and 3. The first gives rise to a D_0 i.r, which, being a singlet, has no orbital degeneracy to be resolved. Nor does it have any associated orbital moment. The five states in $L = 2$ can be used as a basis for a D_2 representation in spherical symmetry, so if the symmetry is lowered it can be expected that the fivefold degeneracy will be lifted to give, at most, five orbital levels. In cubic (octahedral) symmetry there is sufficient symmetry left for some degeneracy to remain. The D_2 i.r. of the rotation group decomposes into a sum of two i.r.'s of the cubic group, $E + T_2$. (Each rotation of the cubic group occurs in the full rotation group.) Its character in D_2 is obtained by substituting the appropriate angle in eqn. (2.14). The character of a representative element in each class of the cubic group is then readily found. It is the set of these which leads to $E + T_2$. So if a splitting does occur it will be into a doublet of E symmetry and a triplet of T_2 symmetry. $L = 3$ gives a D_3 i.r. in spherical symmetry which, in cubic symmetry decomposes into $A_2 + T_1 + T_2$. So the seven levels can be expected to split into a singlet and two triplets. It is typical of group theory that while it gives predictions of this type, which are very useful, it gives no information about the arrangement or magnitudes of the energy level splittings. These can only come from a more detailed study. It also helps in another sense, for it suggests that it will be possible to diagonalize algebraically the matrices that arise in first-order perturbation theory. This is no mean help, for with $L = 2$ the matrix is 5×5 and with $L = 3$ it is 7×7.

§ 4.2 A $(d)^1$ Ion

To illustrate the procedure it is convenient to begin with Ti^{3+}, which has one $3d$ electron ($l = 2$) outside the $(1s)$, $(2s)$, $(2p)$, $(3s)$, and $(3p)$ filled shells. The ion is assumed to be placed at the center of a regular octahedron formed by six identical negative charges that have been attracted to the positive ion. The general Hamiltonian for a many-electron system, neglecting all spin-dependent interactions, takes the form

$$H = \sum_i \frac{\mathbf{p}_i^2}{2m} + \frac{1}{2} \sum_{i \neq j} \frac{e^2}{r_{ij}} + \sum_i U(x_i, y_i, z_i), \qquad (4.1)$$

where a crystal field energy, a summation over all electrons of a single-electron energy $U(x, y, z)$, has been added to the largest energy terms in the free ion Hamiltonian. With the nucleus of the Ti^{3+} ion as origin, the energy of an electron at (x, y, z) due to a charge $-q$ at $(0, 0, a)$ is

$$\frac{eq}{[x^2 + y^2 + (z - a)^2]^{1/2}}, \qquad (4.2)$$

where e is the magnitude of the charge on the electron. This expression can be expanded, on the assumption that the $-q$ charge is outside the Ti^{3+} ion, so that $a > r$, where $r^2 = x^2 + y^2 + z^2$. The simplest way is to use the generating function for the Legendre polynomials:

$$\frac{1}{[1 - 2tu + t^2]^{1/2}} = \sum_{n=0}^{\infty} t^n u^{-(n+1)} P_n(u), \qquad (4.3)$$

which is valid when $|t| < 1$, $|u| \leq 1$, and $|t| < |u|$. The expansion of (4.2) is obtained by setting $t = r/a$ and $u = z/r$. Usually only the first few Legendre polynomials are needed:

$$P_0(u) = 1, \quad P_1(u) = u, \quad P_2(u) = (3u^2 - 1)/2,$$
$$P_3(u) = (5u^3 - 3u)/2, \quad P_4(u) = (35u^4 - 30u^2 + 3)/8.$$

On substituting for t and u and multiplying by an appropriate power of r they become homogeneous functions of x, y, and z, so since potential functions satisfy Laplace's equation they become multiples of the $r^l Y_l^0$ functions. The energy of an electron in the potential due to the charge at $(0, 0, a)$ thus becomes

$$\frac{qe}{a} \left[1 + \frac{z}{a} + \frac{1}{2a^2}(3z^2 - r^2) + \frac{1}{2a^3}(5z^3 - 3zr^2) \right.$$
$$\left. + \frac{1}{8a^4}(35z^4 - 30r^2 z^2 + 3r^4) + \ldots \right]. \qquad (4.4)$$

The energy from any other neighbor is readily obtained by the appropriate interchange of x, y, and z and the total energy from all six neighbors, charges $-q$ at $(\pm a, 0, 0)$, $(0, \pm a, 0)$ and $(0, 0, \pm a)$, can either be obtained by adding the separate energies or by using the cubic symmetry. It is

$$U(x, y, z) = \frac{6qe}{a} \left[1 + \left(\frac{35}{4a^4} \right) \left(x^4 + y^4 + z^4 - \frac{3}{5} r^4 \right) + \ldots \right]. \qquad (4.5)$$

The first term, being independent of x, y, and z, can be dropped as it gives an energy shift, which is common to all energy levels and so is of no interest. In *first*-order perturbation theory the free ion bras and kets of the ground terms span D_2 i.r.'s, and since $D_2 \times D_2 = D_0 + D_1 + D_2 + D_3 + D_4$, it follows that the matrix element, within this set, of any operator that transforms as a D_n with n greater than 4 will be zero. There is therefore no point in retaining the terms in the expansion other than those in $[x^4 + y^4 + \ldots]$. (The general result has been used that $\Gamma_i \times \Gamma_j$ only contains the invariant representation, A_1, if Γ_i is the adjoint of Γ_j.)

To continue, it is convenient to assume, for the moment, that the electrons in closed shells can be neglected, for this removes the need to consider the Coulomb interactions between the electrons and reduces the problem to that of a single electron. As the model Hamiltonian contains no spin variables, the next step is to set up the 5×5 matrix of the crystal field energy using any convenient basis set of $(3d)$ states. In the evaluation of the matrix elements the integrations over polar variables cannot be completed, because the radial parts of the (3d) functions, which are common to all the wave functions, are usually unknown. The angular parts of the integrations, on the other hand, involve standard functions and so present no problems. The result is a matrix in which all the elements appear as multiples of $\langle r^4 \rangle$, the mean of r^4 taken over the radial part of the wave function. With a single d electron the calculation is basically straightforward, though the evaluation of a number of angular integrals is somewhat tedious. As doing this has now been replaced by a simpler method, the straightforward way will not be pursued, for the more powerful method will be described shortly.

The above method becomes even more tedious when the number of electrons is increased. Nevertheless, it was successfully developed in the 1930s and applied to all the iron group ions. There were also some tentative attempts to extend it to the rare earth ions, though doing so presented additional problems. The electrons are in $(4f)$ orbitals, which means that the expansion of the crystal field potential can only be truncated at the sixth order in (x, y, z), for it is only beyond this that all the matrix elements needed in first order of perturbation theory vanish. Then it seemed probable that the crystal field energy would be smaller than the spin-orbit interaction, in which case it would need to be taken into account after the spin-orbit interaction. So it would be a perturbation of the J multiplets of the $(4f)^n$ ions rather than of the L-S terms of the $(3d)^n$ ions. Compared with the iron group ions, the structure of the states in multiplets is much more complicated, and particularly so when the number of electrons ranges from one to thirteen. And finally there was much less certainty about the symmetry of the local environment. So a choice was made, octahedral symmetry with small distortions, a choice which is now known to be relatively uncommon for rare earth ions. Nevertheless, quite good agreement with such experimental results as were available was obtained by choosing the "right" values for the crystal field coefficients. (The main experimental results then available were, for both the iron and rare earth ions, a mixture of temperature-dependent paramagnetic susceptibilities and rather similar results on the rotation of polarized light as it passed through a crystal, taken over a limited range of temperature.)

The theory was subjected to much more demanding tests when the technique of electron paramagnetic resonance (EPR) was introduced in the late 1940s. This made it possible to study individual magnetic ions in crystals in considerable detail, and it quickly confirmed the power of crystal field theory when applied to iron group ions, particularly when the extensions needed to take account of small departures from octahedral symmetry and spin-dependent interactions were made. The consequence was that the success of crystal field theory in the 1930s was further confirmed, and in much more detail, some fifteen or so years later.

In due course EPR was used to study the rare earth ions, with the natural assumption that crystal field theory would be equally successful with these. Indeed it was, but not as a simple extension of the iron group theory, for the typical surroundings of rare earth ions in crystals are quite different from those of typical iron group ions. The crystal field is usually far from being that of a distorted octahedron. It was therefore assumed that in a typical crystal there would be no dominant term in the expansion of the crystal field energy. It would have the general form

$$U(x, y, z) = \sum_{n,m} A_n^m U_n^m (x, y, z), \qquad (4.6)$$

where A_n^m is the coefficient of a U_n^m that satisfies Laplace's equation and transforms in the same way as $r^n Y_n^m$. (In first order of perturbation theory the expansion could be simplified by dropping all U_{nm} with n odd, and all with even n where n is greater than 6, because their matrix elements vanish.) With no dominant A_n^m and so many coefficients, crystal field theory becomes much more complicated. Indeed, in a private communication to your author (February 1951) Professor Van Vleck expressed his doubt about the possibility of using it to explain some of the early results on cerium ethyl sulfate, obtained by the study of the Faraday effect, the rotation of the plane of polarization of light on passing through a magnetic crystal. However, by that date EPR results had become available on cerium and some other rare earth ethyl sulfates which, when combined with a new technique for dealing with the theoretical complications, had led to an explanation of their magnetic properties using crystal field theory, which once again confirmed its power.

The new theoretical method was based on the Wigner-Eckart theorem and amounted to replacing an expression such as that in (4.6) by an expression in angular momentum operators. It rapidly replaced the previous technique and is still in use, though there have been a sequence of changes and some diversity in notation, which has led to a degree of confusion between the meanings of identical symbols used by different authors (see Rudowicz, 1987).

§ 4.3 Operator Equivalents

The crystal field energy, U, is a potential energy summed over all electrons and, in first order of perturbation theory, it is usually necessary to set up its matrix using states which span a degenerate manifold and which transform as an i.r. of the rotation group. For an iron group element this would be a D_L, where L is the orbital momentum of the lowest term, and for a rare earth ion it would be a D_J, where J is the total angular momentum of the lowest multiplet. The expansion of the crystal field energy is already a sum of parts, each of which is a basis element of an i.r. of the rotation group, so all the elements in the matrix are sums of matrix elements, each of which consists of a bra, an operator, and a ket, all of which transform as i.r.'s. The Wigner-Eckart theorem can therefore be used to replace the operator by a reduced matrix element multiplying an angular momentum operator that transforms in the same way as the operator. The procedure has a number of advantages, for once the reduced element has been found for the operator with a given n it can be used for all the other operators that have the same n. Also, the magnitudes of the angular momentum operators are easily found for states of a given L or J.

One such operator, leaving out the scalar coefficient with which it is multiplied in the full expression for the crystal field energy, is $r^2 Y_2^0$, or $(1/2)[3z^2 - r^2]$. It can be replaced by $(1/2)\alpha[3l_z^2 - l(l + 1)]$, where α is a reduced matrix element, if a single electron with angular momentum l is being considered. α is then most simply obtained by comparing the matrix elements of $([3z^2 - r^2]$ and $\alpha[3l_z^2 - l(l + 1)]$ for the case when $m_l = 0$. In the second, or angular momentum form, its value with $l = 2$ is -6α. In the first form it is $\langle 0|3z^2 - r^2|0\rangle$, an integral taken between states with $m = 0$. Its integrand is an expression that is a product of a simple angular form and a radial function containing an unknown part, coming from the radial parts of the l wave functions. The former is readily evaluated and the latter is simply replaced by $\langle r^2 \rangle$, to give $\alpha = (-2/21)$ $\langle r^2 \rangle$. For a many-electron system with orbital angular moment L, the operator $(1/2)[3z^2 - r^2]$ is replaced by a summation over all electrons, and capital letters replace the lower-case letters in the angular momentum equivalent, so that it becomes $(1/2)\alpha[3L_z^2 - L(L + 1)]$. It again becomes necessary to determine the α, which can be expected to take a different value. (In more recent work there is a also a change in notation. Reduced matrix elements are usually written as if they are modified matrix elements. That is, $\alpha \langle r^2 \rangle$ may well be written either as $\langle L||\alpha||L\rangle\langle r^2 \rangle$ or simply as $\langle L||\alpha||L\rangle$.)

By this procedure the crystal field Hamiltonian can be replaced by an expression in angular momentum variables. This is really only valid if the intention is to stop at first order in perturbation theory, a point which is sometimes overlooked, particularly in papers that omit the stage of writing the crystal field out as in (4.6) and go immediately to a form in angular momentum variables. What is even less satisfactory is to find such a form described as a crystal field.

The value of the replacement can be seen if the matrix elements of $(x^2 - y^2)$, summed over all electrons of a many-electron system, are needed within a manifold of L. All that is needed is to replace the summation by $\langle L||\alpha||L\rangle(L_x^2 - L_y^2)$ or the rather more convenient form $(1/2)\langle L||\alpha||L\rangle(L_+^2 + L_-^2)$. It is then immediately seen that its only nonvanishing matrix elements are between states which have ΔM's, differences in M values, of ± 2 and that their values, as multiples of $\langle L||\alpha||L\rangle$, can be obtained using just the properties of angular momentum operators. The number of integrations is much reduced, for the value of $\langle L||\alpha||L\rangle$ is common to all the matrix elements that are symmetry related through an i.r.

There are some further points that should be mentioned. In the above examples, it is easy to see which angular momentum expressions should replace the chosen potentials, which is not the case with some of the more complicated ones. One that regularly crops up in octahedral symmetry is

$$x^4 + y^4 + z^4 - (3/5)r^4.$$

In the first three terms and for a single electron, x can be replaced by l_x, y by l_y, and z by l_z, but then one arrives at r^4, which contains $x^2 y^2$. There is then a problem, for while x and y commute l_x and l_y do not, so the replacement needs to be symmetrized. This complicates the replacement. There is no need to go into the details, though, for all the required replacements were determined some while ago (Stevens, 1952) and all the nonzero matrix elements were listed. They are repeated in a number of texts, such as (A-B), Hutchings (1964), and Craik (1995). The only point that needs watching is that different authors, in using expressions such as that in (4.6), have used different definitions for the U_n^m's and so have arrived at different angular momentum replacements for them. So while O_n^m is now regularly used for an angular momentum operator that transforms in the same way as $r^n Y_n^m$ there can be subtle variations in the normalization factors which relate the two. The best advice as to how to thread one's way through what has become a rather untidy development is to refer to a sequence of papers by Rudowicz (see §4.12). [In (A-B), for example, O_4^4 is $(1/2)[L_+^4 + L_-^4]$, which is not the same as the O_4^4 now commonly used, which contains only L_+^4.]

There is a useful property of all the angular momentum operators, that they all have zero trace. A proof is as follows. The only nonzero matrix elements of operators that correspond to having $m \neq 0$ in the $r^l Y_l^m$'s are off diagonal in a basis that uses the eigenstates of l_z, for all these operators involve powers of l_+ or l_-. So the trace of each operator, the sum of its diagonal elements, is zero. All that is then needed is to show that the operators with $m = 0$ also have zero traces. The operators in a basis for D_l range over all allowed values for m. Further, the trace of any such operator is invariant under a change of basis. Now there must be some rotations that take a given member of the basis into a linear combination containing an operator with $m = 0$, for otherwise the representation would not be irreducible, so it follows that the operator with $m = 0$ also has zero trace.

Returning to the Ti^{3+} example, the presence of a single $3d$ electron outside the closed shell structure means that there is only one term in the ground configuration and that it has $L = 2$. Also, the only term left in the crystal field potential in cubic symmetry is a multiple of $[x^4 + y^4 + z^4 - (3/5)r^4]$. The bras and kets span D_2 i.r.'s and the potential is a basis element for a D_4 i.r., so all is set for invoking the Wigner-Eckart theorem. In first order of perturbation theory $[x^4 + y^4 + z^4 - (3/5)r^4]$ can therefore be replaced by an equivalent operator, which has been found to have the form

$$\langle L||\beta||L\rangle[O_4^0 + 5(O_4^4 + O_4^{-4})],$$

where

$$O_4^0 = 35L_z^4 - 30L(L+1)L_z^2 + 25L_z^2 - 6L(L+1) + 3L^2(L+1)^2$$

and

$$[O_4^4 + O_4^{-4}] = (1/2)[L_+^4 + L_-^4],$$

with $L = 2$. The reduced matrix element, $\langle L||\beta||L\rangle$, is found by taking a convenient state and equating two equivalent expressions. O_4^0 is the equivalent of $35z^4 - 30r^2z^2 + 3r^4$, so

$$\langle M = 0|\Sigma(35z^4 - 30r^2z^2 + 3r^4)_i|M = 0\rangle$$

can be set equal to $72\langle L||\beta||L\rangle$, where 72 is the numerical value of O_4^0 when $L = 2$ and $M = 0$.

The above replacement can be made when Ti^{3+} is assumed to have just the single d electron or equally when it also has the electrons in closed shells. If there is to be a significant difference it will occur in the determination of $\langle L||\beta||L\rangle$. For the single-electron case, it is necessary to have the $L = 2$, $M = 0$ state expressed as a simple $l = 2$, $m = 0$ hydrogenic state and for the many-electron case, as an $L = 2$, $M = 0$ determinantal state with all the filled shell electrons described. (While it is not essential to choose $M = 0$ it is a convenient choice because in the many-electron case the $M = 0$ state is a single determinant.) For it the matrix element reduces to a sum of matrix elements, each of which could occur in a one-electron system. The electrons in closed shells contribute a sum of integrals which can be recognized as a sum of the traces of O_4^0 for each of the l values of the inner shell electrons. As they are all zero, the calculation for the many-electron Ti^{3+} ion reduces to that for a single d electron. So the reduced matrix element, $\langle L||\beta||L\rangle$, is the same for both cases. The only requirement left is the evaluation of

$$\langle m = 0|(35z^4 - 30r^2z^2 + 3r^4)|m = 0\rangle$$

for a single $3d$ electron. It comes out to be $(16/7)\langle r^4\rangle$, so $\langle L||\beta||L\rangle = (2/63)\langle r^4\rangle$, a positive quantity. The final step in setting up the equivalence is to replace $[O_4^4 + O_4^{-4}]$ by $(1/2)\langle L||\beta||L\rangle[L_+^4 + L_-^4]$.

The determination of the first-order eigenstates and eigenvalues now becomes a simple matter, for if the angular momentum operator $O_4^0 + 5[O_4^4 + O_4^{-4}]$ is applied to the ket with $M = 0$, the O_4^0 part will leave the M value unaltered, the O_4^4 part will change the $M = 0$ to $M = 4$, and O_4^{-4} will change it to -4. However, there are no kets in $L = 2$ that have $M = \pm 4$ so applying O_4^4 and O_4^{-4} gives zero. (When the calculation is carried out step by step each application of an L_+ or an L_- operator introduces a numerical factor. In the present example some of the factors are zero.) It follows that $|0\rangle$ is an eigenstate with eigenvalue $72\langle L||\beta||L\rangle$. A similar argument shows that $|1\rangle$ and $|-1\rangle$ are eigenstates with the same eigenvalue, $-48\langle L||\beta||L\rangle$. The remaining two eigenstates are the symmetric and antisymmetric combinations $[|2\rangle \pm |-2\rangle]/\sqrt{2}$, for L_+^4 applied to $|-2\rangle$ gives a multiple of $|2\rangle$ and L_-^4 applied to $|2\rangle$ give the same multiple of $|-2\rangle$. The combination with the $+$ sign has eigenvalue $72\langle L||\beta||L\rangle$ and that with the $-$ sign has eigenvalue $-48\langle L||\beta||L\rangle$. Thus without needing to evaluate $\langle r^4 \rangle$ the zeroth-order eigenstates have been found and, as can be expected from group theory, they give a doublet and a triplet. The new result is that in units of $qe/8a^5$, the doublet is at $72\langle L||\beta||L\rangle$ and above the triplet, which is at $-48\langle L||\beta||L\rangle$. With a doublet having a factor of 72 and a triplet having -48, the mean is zero, as it should be, for the trace of a crystal field operator.

The result can be given a physical interpretation. An electron in the field of an octahedron of negative charges, each of the same magnitude, $-q$, will have its lowest energy when it is as far away as possible from the octahedral positions. There are, however, constraints on its spatial distribution, particularly in first order of perturbation theory where the eigenstates are constructed from $3d$ orbitals. These can be combined so as to form bases for the i.r.'s of the cubic group which, when written in Cartesian coordinates, are $\{xy, yz, zx\}$-like for the T_2 i.r. and $\{x^2 - y^2, 3z^2 - r^2\}$-like for E. Those in T_2 are zero along the axes and largest between the axes, whereas those of E are largest on the axes and zero in between. It is therefore not surprising to find that the T_2 family has a lower energy than the E family. (There is another point, that eigenfunctions are necessarily orthogonal. A linear combination that gives a low-lying eigenstate restricts the available combinations for the others, which usually means that other eigenstates seem to be raised in energy. In diagonalizing a matrix the mean of the trace describes an energy displacement that is common to all its eigenvalues, so it can usually be neglected. The trace is then zero. So if the diagonalization lowers one diagonal element it will raise some of the others.)

It has been convenient to assume that the surroundings of the Ti^{+3} ion have octahedral symmetry, for with the equivalent operators the treatment of $L = 2$ is very simple, as is also the treatment of $L = 3$. However, it is usually found that the octahedron of six neighbors is slightly distorted, so that the symmetry is lowered. Two types of distortion are commonly found, a tetragonal distortion which leaves a fourfold axis of symmetry and a trigonal distortion which leaves a threefold axis. In both cases it is best to take the axis of distortion as Oz. For a tetragonal distortion the crystal field, when written in the equivalent operator formalism appropriate to first-order perturbation theory, has a systematic form:

$$B_2^0 O_2^0 + B_4^0 O_4^0 + B_4^4 O_4^4 + B_4^{-4} O_4^{-4}, \tag{4.7}$$

where the B_n^m's coefficients have the dimensions of energy and the O_n^m are equivalent operators in L. Those with odd values of n have been omitted because they have zero matrix elements in first order. (They reverse on inversion, the substitution $x \rightarrow -x$, $y \rightarrow -y$, $z \rightarrow -z$.) In this example and many others, inversion is a symmetry operation of the crystal field, a property that is not essential in showing the above. In evaluating a matrix element between two states of the same l the integrand always contains a product of a ψ and a ψ^* which is invariant under inversion because all states of the same l are either unaltered (l even) or reversed (l odd) under inversion. So if the V in $\psi V^* \psi$ is reversed under inversion, the matrix element vanishes. Other operators, such as O_4^2, are omitted for another reason. The presence of the superscript 2 indicates that it comes from a potential that contains $(x + iy)^2$, which has the property that if the axes are rotated about Oz through $\pi/2$ the transformation $x \rightarrow y$ and $y \rightarrow -x$ is induced. This reverses $(x + iy)^2$, which is incompatible with tetragonal symmetry. Only combinations of x and y in the forms $(x \pm iy)^4$ can occur, and only then in such a way that the potential is real. Thus only O_2^0 and O_4^0, which do not contain x and y, and O_4^4 and O_4^{-4}, with identical coefficients, occur in the effective crystal field. [In some of the papers dealing with tetragonal symmetry, the Ox and Oy axes are rotated about Oz through $\pi/4$. While this makes no difference to the $m = 0$ parts of (4.6) it leads to sign reversals for B_4^4 and B_4^{-4}.]

With $L = 2$ the states $|0\rangle$, $|1\rangle$, and $|-1\rangle$ remain as eigenstates, with $|1\rangle$ and $|-1\rangle$ staying degenerate. Similarly, $|2\rangle$ and $|-2\rangle$ are again mixed into the symmetric and antisymmetric combinations $[|2\rangle \pm |-2\rangle]/\sqrt{2}$, for the two diagonal matrix elements are equal. These two states were not degenerate in octahedral symmetry and there is no reason to suppose they will be in tetragonal symmetry. Previously, though, the combination with the minus sign was coincident with $|1\rangle$ and $|-1\rangle$. This will no longer occur, so the pattern in octahedral symmetry of two energy levels, one doubly and the other triply degenerate, will be changed into two singlets from the doublet and a singlet and a doublet from the triplet.

The doublet will be split if there is an extra distortion that adds a perturbation with a matrix element between $|1\rangle$ and $|-1\rangle$. This requires an operator that connects states with $\Delta M = 2$. It can only come from a potential containing a real combination of $(x + iy)^2$ and $(x - iy)^2$. Such potentials are present when there is a twofold axis of symmetry and also when the symmetry is further reduced. (High symmetry has the effect of eliminating expressions which, on general arguments, might be expected to be present. The lower the symmetry, the more nonzero potentials there are.) So even if Oz is still a twofold axis of symmetry, the only remaining degeneracy will be lifted.

In trigonal symmetry the crystal field operator in its angular momentum form is

$$B_2^0 O_2^0 + B_4^0 O_4^0 + B_4^3 O_4^3 + B_4^{-3} O_4^{-3}. \tag{4.8}$$

With ΔM's of 0 and 3, $|1\rangle$ can only mix with $|-2\rangle$ and $|-1\rangle$ with $|2\rangle$ and there is no state that can mix with $|0\rangle$. There will therefore be two doublets and a singlet, for the submatrix based on $|1\rangle$ and $|-2\rangle$ is identical with the submatrix based on $|-1\rangle$ and $|2\rangle$. Thus under a trigonal distortion the triplet in octahedral symmetry will split into a singlet and a doublet and the doublet will remain so. Any further lowering of symmetry will split both doublets.

The next step is to consider an ion with more than one electron in the $(3d)$ shell. With equivalent operators the changes needed are trivial. The ground state "terms" of all the $(3d)$ ions conform to Hund's rule, and in the sequence $(d)^1$, $(d)^2$, etc. they are respectively ^2D, ^3F, ^4F, ^5D, ^6S, ^5D, ^4F, ^3F, ^2D, so they are symmetric about the middle $(3d)^5$ configuration and contain only $L = 0, 2$, and 3. With $L = 0$ the crystal field is effectively a constant, for there is no orbital degeneracy to resolve. For the two other values the crystal field is simply replaced by

$$\sum_{n,m} B_n^m O_n^m (L), \tag{4.9}$$

where the O_n^m are equivalent operators with the appropriate L in place of l. The B_n^m are again often referred to as crystal field parameters. As for Ti^{3+}, they contain the detailed information about the crystal field, the mean values of powers of r, and reduced matrix elements. In each case the most convenient way to find the reduced matrix elements is to take the state in L that has $M_L = 0$ and write it in its Slater determinantal form, for, in the ground terms, it is a single determinant. The matrix element for an equivalent operator that has a definite value for n and $m = 0$ reduces to a sum of diagonal elements of the corresponding one-electron crystal field operator, which is easily reduced to an expression in the single-electron $\langle l||\alpha||l\rangle$ or $\langle l||\beta||l\rangle$ values. (α is used for quadratic potentials, β for quartic, and so on.) For those configurations that have $L = 2$ as the lowest term, $\langle l||\alpha||l\rangle$ and $\langle l||\beta||l\rangle$ have the same values for $(3d)^1$ and $(3d)^6$ and reversed values for $(3d)^4$ and $(3d)^6$. For the $L = 3$ examples the values are $\langle l||\alpha||l\rangle = -2/105$ and $\langle l||\beta||l\rangle = -2/315$ for $(3d)^2$ and $(3d)^7$, with reversed signs for $(3d)^3$ and $(3d)^8$. These alternations in sign are of no significance when it comes to choosing values for the B_n^m to fit observations. They are important, though, if an attempt is made to compare the energy levels of two different ions with the same L that are thought to be in the same crystal field.

The crystal field given in (4.4) was introduced initially because the subsequent theory gives the eigenvalues as expressions in $(qe/8a^5)$, $\langle l||\beta||l\rangle$ and $\langle r^4\rangle$, leaving only q and $\langle r^4\rangle$ to be estimated. Taking what seemed reasonable values for these then produced crystal field splittings not unlike those which appeared to be present. Subsequent "improvements," such as taking account of the charge distributions on neighboring ions and nonorthogonality between their wave functions and those of the central ion, tended to spoil the rather satisfactory agreement between theory and experiment, so nowadays it is more usual to take the crystal field directly in an angular momentum form, such as that given in (4.9), and to choose the B_n^m coefficients to fit experimental observations. While this has had the effect of obscuring the details of the neighboring charge distributions and, possibly, extending the concept of a crystal field, it does not affect the argument that on moving from an ion with one set of reduced matrix elements to an ion with a reversed set, in what is substantially the same environment, the energy level pattern should be reversed. As this is commonly observed, the validity of crystal field theory can be said to be supported.

For any iron group ion, the maximum size of the matrix that needs to be diagonalized in first-order perturbation theory is 7×7, so no difficulties are presented if the elements are given as numbers. All that is necessary is to make numerical choices for the B_n^m and have available the matrix elements of the relevant angular momentum operators. The latter can be found in various articles and books [e.g., table 17 of (A-B) and the review by Hutchings (1964)]. More recently, there has been a move away from them, to another set based on the spherical tensor operators introduced by Racah (1942). The two sets of operators are closely related, the main difference being that the two have different normalization factors (Judd, 1963). (This is partially hidden when the Racah operators are expressed via their matrix elements in a particular representation.)

There is a feature of the $L = 3$ case with cubic symmetry that is of some interest because the singlet, triplet, triplet pattern comes out to have a specific ordering, A_2, T_2, T_1, with the ratio of the separations between the first and last pairs being 5:4. Such a prediction is surprising because there is no obvious reason why symmetry should predict it. In fact, it does not. What has happened is that an extra feature has been slipped in. The d-like nature of the one-electron wave functions has resulted in all the matrix elements associated with potentials with n greater than 4 being zero and this has reduced the operator equivalent to that of a single parameter multiplying $[O_4^0 + 5(O_4^4 + O_4^{-4})]$. It is this expression that has produced the 5:4 ratio. Had the $3d$ restriction not been imposed, there would have been many more crystal field terms of cubic symmetry, such as $(x^6 + y^6 + \cdots)$ and the equivalent operator form would have contained many more parameters. The 5:4 ratio would have disappeared. The observation of such a pattern for a number of ions with $L = 3$ has been used to support the assumption that the one-electron wave functions are close to being $3d$-like.

Combining the equivalent operator technique with the Bloch form of perturbation theory has a number of attractions, for to all orders in perturbation theory the operator obtained is bordered by P_0's. So it acts in the same Hilbert space as that used in zeroth order. Also, any finite group has a finite number of classes and therefore a finite number of i.r.'s. It follows that the effective Hamiltonian, whether it is produced in first- or infinite-order perturbation theory, can be written as a finite sum of operators, each of which belongs to some basis of an i.r. of the symmetry group. (Suppose one takes a particular operator, say L_x^p, for a particular L. Applying all the operations of the symmetry group to it will eventually produce a finite number of operators that can be used to define a basis for a representation of the group. This can then be decomposed into i.r.'s, so providing a means whereby the initial operator, L_x^p, can be written as a sum of operators, each of which belongs to the basis set of some i.r.) Each one can therefore be replaced by a reduced matrix element times an equivalent operator. The number of equivalent operators required is no more than the sum of the characters in the classes of E, the unit operator, and the number of reduced matrix elements is no more than the number of classes.

All this information is in the table of group characters, as can readily be seen. If a particular i.r. has a basis consisting of q equivalent operators, then all the operators in the expansion that transform in the same way as an element, A, in this basis, can be lumped together and replaced by a reduced matrix element times the operator for A. With q such equivalent operators in this i.r., the result will be an expression in q equivalent operators, each of which has the same reduced matrix element. To span all possible i.r.'s will therefore require as many reduced matrix elements as there are classes. Also, each matrix in an i.r. based on p elements is a $p \times p$ matrix, so to determine p it is only necessary to look at the character of the unit operator, for this is a unit matrix. The number of equivalent operators needed to describe the effective Hamiltonian is therefore equal to the sum, over all the i.r.'s, of the characters of E, and the number of reduced matrix elements is equal to the number of classes.

§ 4.4 Electron-Hole Symmetry

The reason for the variations in sign of the reduced matrix elements on going between similar terms for different ions is easily seen by an argument based on replacing electrons by "holes." Any Slater determinant describing a system with n $(3d)$ electrons can be related to a determinant of a system with $(10 - n)$ $(3d)$ electrons by specifying the orbitals that are not occupied rather than those that are. As an example, the state which has the maximum total M_L and M_S in the ground term for a $(3d)^3$ configuration can be compared with the similar state of $(3d)^7$. That of the former can be written as $\{2, +; 1, +; 0+\}$, and that of the latter as $\{2, +; 1, +; 0, +; -1, +; -2; +; 2, -; 1, -\}$. For the first the expectation value of any equivalent angular momentum operator will take the form $\langle 2|U|2 \rangle + \langle 1|U|1 \rangle + \langle 0|U|0 \rangle$, where U is an expression in l, and the corresponding expression for the $(3d)^7$ system will be a sum of seven matrix elements, of which five can be grouped together into

$$\sum_{m=-2}^{m=2} \langle m|U|m \rangle, \qquad (4.10)$$

which equals zero, for it is the trace of U. Using the same result the remaining matrix elements, $\langle -2|U| -2\rangle + \langle -1|U| -1\rangle$, give a value that is the negative of that found for the three-electron case. Thus the seven-electron system has crystal field splittings that are opposite in sign to those of the three-electron system. This is just what is to be expected if the seven-electron system is regarded as composed of three "holes," each with a positive charge. Such a symmetry between electrons and holes can be used in a variety of ways. For example, in first-order perturbation theory the term structure of a $(d)^n$ configuration will be the same as that of a $(d)^{(10-n)}$ configuration, because replacing electrons by holes does not change the Coulomb interactions between charges. (There will be differences in magnitudes, because although the two ions are related they are not identical.) In particular, both configurations will have the same lowest term. If this is then split by a cubic crystal field the pattern for $(d)^n$ will be inverted with respect to that of $(d)^{(10-n)}$. Using similar arguments it can be shown that if the one-electron spin-orbit interactions in the two ions are replaced, in the ground term, by an equivalent operator $\lambda \mathbf{L} \cdot \mathbf{S}$, the reduced matrix element, λ, has opposite signs for the two ions.

In first order of perturbation theory an ion having five $(3d)$ electrons can alternatively be regarded as an ion that has five holes. The electron-hole substitution would then seem to imply that first-order crystal field splittings would be inverted when the hole description is used, which is obviously nonphysical. The only conclusion must be that there are no crystal field splittings of a $(3d)^5$ configuration in first-order perturbation theory. (Nor can there be any first-order splittings due to spin-orbit interactions.) In fact, experiment has shown that there usually are splittings of the sixfold degeneracy associated with S-state ions, but of a much smaller magnitude than that commonly associated with crystal fields. In consequence, there has been a good deal of interest (and difficulty) in explaining the origin of these splittings. (The electron-hole duality only holds for Slater determinants which are constructed from states in the same configuration. Once a perturbation treatment introduces excited states in which electrons have been moved from one electronic shell to another, the duality breaks down.)

§ 4.5 The Quenching of the Spin-Orbit Coupling

So far all spin-dependent interactions have been omitted from the Hamiltonian, and this has resulted in crystal-field split levels being referred to as singlets, doublets, etc., when in fact they have additional degeneracy due to spin. The point has now been reached when these degeneracies and their possible splittings cannot be ignored. The traditional approach in crystal field theory has been to assume that the crystal field theory has produced an isolated orbital ground state and ask, first, whether the spin degeneracy is split in first order of perturbation theory? The answer is arrived at in stages, the first stage being to examine the spin-orbit interactions. The largest of these are of spin–own-orbit character and, using equivalent operators, they can be replaced, within a given term, by $\lambda \mathbf{L} \cdot \mathbf{S}$. Any spin–other-orbit interactions can then be included by modifying λ. The next requirement is the determination of the matrix element of each component of $\mathbf{L}$ within the orbital singlet. This can be shown to be zero, for all singlets. The proof relies on time reversal invariance, for all components of $\mathbf{L}$ are reversed by this operation whereas the ket-bra part of the matrix element is unchanged. The angular momentum is therefore said to be "quenched" in any orbital singlet. An example is furnished by the state $[|2\rangle + |-2\rangle]/\sqrt{2}$, which is an eigenstate of an $L = 2$ ion in a tetragonal crystal field. It is easily seen that the diagonal elements of L_+, L_-, and L_z are all zero.

The quenching in $[|2\rangle + |-2\rangle]/\sqrt{2}$ can be given a physical interpretation by changing to its wave-mechanical form, when the corresponding state has the same angular part as $x^2 - y^2$. In the z plane it has its largest magnitude along the Ox and Oy axes, being zero midway between them. On the other hand, $[|2\rangle - |-2\rangle]/\sqrt{2}$, the state with a reversed sign for $|-2\rangle$, has an xy dependence, so, ignoring its phase, its shape is the same as that of $[|2\rangle + |-2\rangle]/\sqrt{2}$ but rotated through $\pi/4$ about Oz. For a free ion, both states have the same energy, and the charge distribution of the first varies as $[x^2 - y^2]^2$ whereas that of the second varies as $4x^2y^2$. The state $|2\rangle$, which can be represented by $x^2 - y^2 + 2ixy$, has a charge distribution given by its square modulus, which takes the form $(x^2 + y^2)^2$. It is a cylindrically symmetric charge distribution, as is also that of $|-2\rangle$. They differ in that the former has an angular momentum about Oz of 2 units whereas the latter has one of -2 units. In the presence of an external negative charge it can be expected that the rotational motion of an electron in either of these two states will be hindered, and that the lowest-energy state will be one in which its charge distribution has the electron as far away as possible from the hindering negative charges on the axes. This will pick out the $2xy$ combination as the lower-energy state and make $x^2 - y^2$ a higher-energy state, destroying the rotational angular moment in the process. (In higher orders of perturbation theory it is possible to have a partial restoration of the rotational motion, particularly if the splitting due to the hindering is small.)

The next step is usually to ask what effect an applied magnetic field has. It adds two terms to the Hamiltonian, one linear in the field, of the form $\beta \mathbf{B} \cdot (\mathbf{L} + 2\mathbf{S})$, and another that is quadratic in $\mathbf{B} \cdot \mathbf{r}$. The $\beta \mathbf{B} \cdot \mathbf{L}$ part of the former can be dropped because of the quenching, to leave just $2\beta \mathbf{B} \cdot \mathbf{S}$, and the latter, having no spin dependence, simply becomes a quadratic in $\mathbf{B}$. In EPR work it is usually ignored, for it gives a displacement of all the low-lying energy levels. It should not be dropped, though, in a theory of magnetic susceptibility. All the electrons, whether they are in closed or partially filled shells, give contributions to a temperature-independent diamagnetism. (Even the so-called nonmagnetic ions show diamagnetism.) The $2\beta\mathbf{B}\cdot\mathbf{S}$ term is much more interesting for it gives a spin-only Zeeman splitting of the $(2S + 1)$ spin degeneracy, which in turn gives rise to a Curie-like spin-only susceptibility. (In the development of crystal field theory the emergence of a spin-only susceptibility, because of the orbital quenching associated with an orbital singlet, was a major triumph, for it explained why the Zeeman splittings observed in spectroscopy had to be changed to spin-only values to explain the Curie-like observed susceptibilities.)

§ 4.6 The Spin-Hamiltonian

For the $(3d)$ ions the perturbation theory is not usually stopped at first order, and now a complication can occur, for some of the second-order effects may be more important than effects, in first order, from terms that have either been omitted or have given no contributions. For example, a resolution of a first-order degeneracy by a small departure from octahedral symmetry leads to a prediction that a quenching of the spin-orbit interaction should occur, leading to a spin-only paramagnetic susceptibility. If it happens that in second order the spin-orbit interaction gives energy changes that are comparable to or greater than the first-order splittings due to the slight departure from octahedral symmetry, then the two interactions should have been treated on an equal footing.

In an introductory exposition it is not necessary to attempt to examine all possibilities in detail, so for the moment the interactions that are extra to the crystal fields will be restricted to the spin-orbit and Zeeman interactions. It will also be assumed that the first-order effect of the crystal field has been to produce a ground orbital state which is a singlet, and that it is not necessary to take the crystal field to second order. The perturbation will therefore be taken to be

$$\lambda\mathbf{L}\cdot\mathbf{S} + \beta\mathbf{B}\cdot(\mathbf{L} + 2\mathbf{S}),$$

which implies that the excited states will be restricted to states in the same ionic term. (Otherwise, the spin-orbit interaction needs to be modified.)

With an orbital singlet the standard procedure is to use a version of nondegenerate perturbation theory in which the $\mathbf{S}$ operators are simply regarded as noncommuting operators, in which case the second-order correction takes the form

$$\sum_n \frac{|\langle 0|\lambda\mathbf{L}\cdot\mathbf{S} + \beta\mathbf{B}\cdot(\mathbf{L} + 2\mathbf{S})|n\rangle|^2}{E_0 - E_n}, \tag{4.11}$$

where $|0\rangle$ is the ground orbital state, with energy E_0, and $|n\rangle$ ranges over the excited orbital states, which have energies E_n. The term $2\beta\mathbf{B}\cdot\mathbf{S}$ can be dropped because it has no matrix elements between different orbital states, and with the matrix elements of **L** having been determined, the first- and second-order terms, taken together, give an effective operator of the form

$$\mathbf{S} \cdot D \cdot \mathbf{S} + \beta \mathbf{B} \cdot g \cdot \mathbf{S} - \beta^2 \mathbf{B} \cdot \Lambda \cdot \mathbf{B}, \qquad (4.12)$$

where g, D, and Λ are tensors. Such a form is known as a spin-Hamiltonian, for it is an effective Hamiltonian that will give the energy levels into which the $(2S + 1)$-fold degeneracy of the ground orbital level will split in second order of perturbation theory. The part in Λ gives all energy levels the same shift so it is usually ignored in EPR. This should not be done in the theory of the susceptibility. The part in D is present even if **B** is set to zero. If, however, $S = 1/2$ it reduces to a constant and so is of no interest. For other nonzero values of S it gives what are known as the zero-field splittings, splittings of the spin degeneracy that are present even when no magnetic field is applied. In accordance with Kramers' theorem, all zero-field energy levels have an even degeneracy when S is an integer plus one-half. If **B** is nonzero the Zeeman splitting depends in a complicated way on the direction and magnitude of the field, so much so that mapping out the energy level pattern experimentally is a lengthy process, as would be the listing of the results were it not for the spin-Hamiltonian concept. The correct choice of a relatively few constants, those that define g and D, gives a concise formalism which can be used to give a matrix, using any convenient basis set of spin states, the diagonalization of which gives the energy level pattern for any given magnetic field. It is largely for this reason that the concept of a spin-Hamiltonian was so favorably received when it was first introduced, and why it has since been extended far beyond the example just described.

The above spin-Hamiltonian has been obtained using an unusual version of perturbation theory, which, when one is aware of the Bloch scheme, looks relatively clumsy. It is therefore of interest to consider using the Bloch method. The projection operator for the ground level would span the Hilbert space of a spin, **S**, so the effective operator, to any order in perturbation theory, will be an operator in this space. The perturbation contains three vectors, **L**, **S**, and **B**, the first of which is unlike the other two in that wherever it appears it can be replaced by a scalar. So the effective Hamiltonian will be composed of combinations of the components **S** and **B**, which can be arranged into forms that are nonzero within **S** and which transform as basis elements of i.r.'s of the rotation group. This opens the possibility of simply guessing which forms will be present in the effective or spin-Hamiltonian. Going back to (4.12) each spin component of $\beta \mathbf{B} \cdot g \cdot \mathbf{S}$ belongs to a D_1 i.r. and each spin component of $\mathbf{S} \cdot D \cdot \mathbf{S}$ belongs to either D_0 or D_2, so it can be expected that on going beyond second order the values of g and D may be changed slightly, and that more complicated zero-field spin operators and more complicated Zeeman terms may appear. But with the limitation on S that is always present the possibilities are restricted.

At this point it is of interest to consider a situation in which the ground orbital level is degenerate, or nearly so. It would then seem best to define P_0 so that it spans the orbital quasi degeneracy as well as the spin degeneracy. For three closely spaced levels it is then convenient to introduce an effective $L = 1$ to span the orbital quasi degeneracy, and either go through the Bloch perturbation scheme in some detail or simply look for an effective operator in **L**, **S**, and **B** that gives eigenvalues to fit the experimental observations. An orbital quasi degeneracy of 2 can be spanned, in principle, by an effective $L = 1/2$, but as this looks strange the more usual procedure is to invent an effective $T = 1/2$. Then there can be additional interactions in the Hamiltonian that involve one or more nuclear spins, and to cope with this the Hilbert space of P_0 has to be further expanded, by introducing one or more I operators. So one way or another a variety of spin-Hamiltonians have come into use, and where they have come from is not always clear. But what can be relied on is that the number of their eigenstates is equal to the number of low-lying states, and that they have the virtue that if the parameters they contain have been correctly chosen they encompass a mass of experimental information in a highly condensed way.

§ 4.7 Rare Earth Ions

Before going on to more examples of spin-Hamiltonians, it is convenient to consider the crystal field theory of rare earth ions, so that they can be incorporated into the same scheme. The expectation was that, compared with the iron group ions, the crystal fields would be weaker, because the $(4f)$ electrons are deeper in the ion, and the spin-orbit interactions would be stronger, because the magnetic electrons are nearer the nucleus. These expectations have been amply confirmed, and the general assumption is that in the theoretical development the crystal field should be regarded as a perturbation to be applied after account has been taken of the spin-orbit interactions. The free ion theory can therefore be followed rather further, for the role of the crystal field becomes one of splitting the manifolds of total J. For most ions it is only necessary to consider the lowest J manifold, with the value of J being governed by one of Hund's rules, which also emerges from the free ion theory. It states that in the first half of the f shell the ground multiplet has the minimum J value, whereas in the second half it has the maximum value. That is, for the first half of the shell J is equal to $|L - S|$ and for the second half it is equal to $L + S$. For the ions with half-filled shells, those with seven $(4f)$ electrons, Eu^{2+} and Gd^{3+}, $L = 0$ and $J = 7/2$ so no crystal field splittings are to be expected, except that as with the S-state ions in the iron group small zero-field splittings are found. A J of $7/2$ is greater than any L in the ground terms of the $3d$ ions; even so, it is less than the largest value, 8, for a rare earth ion, which occurs in the ground J of $(4f)^{10}$. Another difference is that as the ions are larger than the iron group ions, they usually have more nonmagnetic neighbors, nine or more being common. In consequence, the crystal fields are seldom dominantly octahedral.

At first sight it might seem that the crystal field theory of rare earth ions would be more complicated than that of the iron group ions, because the expansion corresponding to (4.4) can only be truncated beyond the sixth order (D_2 should now be replaced by D_3 in $D_2 \times D_2$, to give $D_0 + D_1$, etc., up to D_6). Also, there is no reason to suppose that any of the terms that are left in the expansion, when those odd in n are dropped, will dominate. The saving grace is that having replaced the crystal field terms by equivalent operators in J, the lowest-lying level is usually either a singlet or a quasi or actual doublet. (The actual doublets occur particularly with ions that have an odd number of $(4f)$ electrons—a consequence of Kramers' theorem.) In the singlets both $\mathbf{L}$ and $\mathbf{S}$ are quenched, so there is no Zeeman splitting to account for, and in the doublets the Zeeman splitting is usually so anisotropic that the scope for choosing pairs of states that will give the observed anisotropies is very restricted. (It can happen that several choices of crystal field parameters will give the same anisotropy. It is then usually necessary to examine higher-lying levels, to distinguish between the choices.) The main complication, at least in the early work, was in the determination of the reduced matrix elements, of which there are now three. However, this only has to be done once, for each ion; all the required values are tabulated in (A-B) and in Hutchings (1964).

§ 4.8 Some Examples of Spin-Hamiltonians

The simplest example of a spin-Hamiltonian occurs, for an iron group ion, when the ground orbital level is nondegenerate and the electron spin is $1/2$. The spin-Hamiltonian is then usually taken to have the form $\mathcal{H}_S = \beta \mathbf{S} \cdot g \cdot \mathbf{B}$, where g is called the g tensor. Choosing its principal axes as coordinate directions, it can be written in the form

$$\mathcal{H}_S = \beta[g_x B_x S_x + g_y B_y S_y + g_z B_z S_z].$$

In many cases these axes bear no relation to the crystallographic axes. Also, the spin-Hamiltonian is incomplete, because there are usually terms quadratic in the components of $\mathbf{B}$ that have no dependence on $\mathbf{S}$. They produce common energy shifts, which are not observed by resonance techniques. For iron group ions the values of g_x, g_y, and g_z are usually close to 2, the free electron value. For the rare earth ions the position is quite different, and there is often considerable anisotropy in the g values. Indeed, in ions with an even number of $(4f)$ electrons and, usually, a twofold quasi degeneracy, some of the principal components of g may be close or even equal to zero. The next simplest case is when there is a nuclear spin of $1/2$. This is incorporated by adding $\mathbf{S} \cdot A \cdot \mathbf{I} - \mu_N g_N \mathbf{I} \cdot \mathbf{B}$, to the spin-Hamiltonian, where $\mathbf{I}$ is the angular momentum operator for a nuclear spin $I = 1/2$. A is an anisotropic hyperfine coupling tensor which, though it is similar to the g tensor, does not usually have its principal axes in the same directions, and g_N is the nuclear g value, an isotropic quantity. (The external magnetic field acts on the nucleus as well as on the electrons.) If the nuclear spin happens to be greater than $1/2$ the value of I in the above expression is adjusted, and since the nucleus usually then has an electric quadrupole moment, a further term of the form $\mathbf{I} \cdot P \cdot \mathbf{I}$ is added to allow for the interaction of the asymmetric electric charge of the nucleus with the charge distribution of the magnetic electrons and with the charges that give rise to the crystal field.

If S is greater than $1/2$ all the above interactions can again occur, S being appropriately adjusted, along with extra interactions, expressions in powers of the components of $\mathbf{S}$. Thus for Cr^{3+}, where the free ion has $L = 3$ and $S = 3/2$, and which is usually to be found in a distorted octahedron that retains an axis of either three- or fourfold symmetry, the zero-field splitting is usually written as $D[3S_z^2 - 15/4]$, the $15/4$ being $S(S+1)$ when $S = 3/2$. (The octahedral crystal field gives a ground orbital singlet.) The coefficients, such as the D, are usually found to be independent of $\mathbf{B}$, though theoretically some dependence is to be expected. Many other examples can be found in (A-B). A glance at these will probably be enough to convince a reader unfamiliar with spin-Hamiltonians that they can be quite complicated expressions. It must be remembered, though, that the experimentally observed properties of magnetic ions in crystals can also be very complicated, in fact much more so. The power of crystal field theory is that it has shown how spin-Hamiltonians can be derived, and how useful they then are in describing observations in a succinct way.

§ 4.9 A Comment about Kramers' Theorem

The theorem is particularly useful when considering EPR, for an ion with an odd number of electrons, for this ion will have the Kramers' degeneracy in all its levels in zero magnetic field. It can therefore be expected to show splittings that are linear in magnetic field for small fields. It is necessary, however, to be careful when applying the theorem to ions in which the electrons are coupled to nuclei, for if the number of electrons is odd and the nuclear spin is also odd, then the combined system does not show the Kramers' degeneracies. The reason is that the electron-nuclear interactions are invariant under the time-reversal symmetry operation, which makes it necessary, in using this operation, to take account of how the nuclear spin states change. The simplest procedure is to begin by forgetting the interactions with the nuclear spin and so be sure of the electronic Kramers' degeneracy if the number of electrons is odd. On then including the electron-nuclear interactions, the Kramers' degeneracy will not be resolved for integer nuclear spins but it probably will be for integers plus one-half. The resolution, however, is likely to be so small that the resonance will still be observable. The opposite situation, in which the number of electrons is even and the nuclear spin is odd, will have the Kramers' degeneracy but usually the prospect for observing EPR is unfavorable. This is because electronic degeneracy is uncommon when the number of electrons is even, so while coupling to a nucleus with a spin that is an integer plus one-half gives the Kramers' degeneracy, the Zeeman splitting is determined by the magnitude of the nuclear moment, and so is much smaller than when it is determined by the electronic moment.

The interactions between internal magnetic moments and interactions with external magnetic fields are treated differently in considering time-reversal operations. The external field is not regarded as reversing, whereas all internal moments do reverse. Internal interactions are therefore time-reversal invariant, whereas interactions with external magnetic fields are not.

§ 4.10 References

Abragam, A., and Bleaney, B. 1970. *Electron Paramagnetic Resonance of Transition Ions* (Clarendon Press: Oxford).

Bethe, H. 1929. *Ann. Physik* 3:133.

Craik, D.J. 1995. *Magnetism: Principles and Applications* (Wiley: Chichester).

Griffith, J.S. 1964. *The Theory of Transition Metal Ions* (University Press: Cambridge).

Hutchings, M.T. 1964. *Solid State Physics* 16:227.

Judd, B.R. 1963. *Operator Techniques in Atomic Spectroscopy* (McGraw-Hill: New York).

Racah, G. 1942. *Phys. Rev.* 62:438.

Rudowicz, C. 1987. *Magn. Res. Rev.* 13:1.

Stevens, K.W.H. 1952. *Proc. Phys. Soc.* A 65:209.

Van Vleck, J.H. 1944. *The Theory of Electric and Magnetic Susceptibilities* (University Press: Oxford), chap. XI.

Van Vleck, J.H. 1950. *Am. J. Phys.* 18:495.

§ 4.11 Further Reading

The book referred to as (A-B), and referenced above, contains many examples of the application of crystal field theory to the interpretation of electron spin resonance (paramagnetic resonance) results. An earlier publication (Griffith, 1964), also referenced above, gives a detailed account of the theory of isolated ions and the changes that occur in a crystal field. There are many later accounts and a number of journals (see the references at the end of Chapter 1.)

There is a problem over listing the replacement angular momentum operators and their matrix elements because of the diversity of their definitions and errors in the tabulated matrix elements. The reader is therefore advised to consult the papers by C. Rudowicz, particularly *J. Phys. C* 18:145 and 3837 (1985a); *J. Magn. Reson.* 63:95 (1985b); *J. Chem. Phys.* 84:5045 (1986); and *Magn. Res. Rev.* 13:1 (1987), for a critical account of the current literature and for proposals for future standardization in the definitions and notation. A step forward would seem to be to adopt his suggestions. (This is not meant to imply that crystal field theory is a viable theory. That the operators emerge from it does not make it acceptable.)

5

Beyond Crystal Field Theory

§ 5.1 Parameter Magnitudes

The previous chapter has shown that the spin-Hamiltonian forms can be derived using crystal field theory and some general assumptions about orders of magnitude, such as that the term separations are small compared with the separations between the configurations but large compared with the crystal field splittings; the spin-orbit splittings are small compared with the term separations; in the iron group ions the crystal field splittings are large compared with the spin-orbit splittings; in the rare earth ions the spin-orbit splittings are large compared with the crystal field splittings. Such assumptions are not by any means enough to give the values of the parameters in the spin-Hamiltonians. These are obtained by experiment, and they are now known very accurately from electron paramagnetic resonance experiments on many magnetic ions present as impurities in otherwise nonmagnetic host crystals. Most of the experiments use frequencies of the order of 10^{10} Hz and a variable external magnetic field of the order of 10^4 G (1 T), which is used to bring energy level differences into resonance. The overall spread of the energy levels described by a spin-Hamiltonian is seldom more than a few cm^{-1} and, for the iron group ions, the next highest set may be of the order of 10,000 cm^{-1} higher if the separation is determined by the cubic part of the crystal field. If it is due to departures from cubic symmetry the separation is not usually so large, being perhaps of the order of 500 cm^{-1}. For rare earth ions the spin-Hamiltonians again describe levels with a spread of the order of 1 cm^{-1}, but now there may be crystal field levels of the order of tens of cm^{-1} away. (The cm^{-1} unit is commonly used in spectroscopy. It is not an energy, but the number of wavelengths of an e.m. wave in 1 cm. So it is proportional to the energy. The frequency of the e.m. wave can be obtained by multiplying its magnitude in cm^{-1} by the velocity, c, of light, in cm per sec. In a similar way any temperature expressed in units of the absolute scale can be related to an energy by multiplying its magnitude by Boltzmann's constant, k. It is useful to know that the energy associated with 1 cm^{-1} is the same as that associated with a temperature of 1.432 K.)

Since most experiments on solids are confined to temperatures below a few thousand degrees, the only energy levels that will be thermally populated are those which are not more than a few thousand cm^{-1} above the ground level. In the study of magnetic ions in crystals, the interest is usually in properties at temperatures much lower than this. For the iron group ions it can therefore be assumed that only the levels described by the spin-Hamiltonian will be thermally populated. For the rare earths the position is more complicated, because there may be several thermally occupied crystal field levels in the temperature range of interest. Since the EPR measurements will probably have been made at liquid helium temperatures, the properties of these higher-lying levels may not be accurately known.

There are other ways of finding out about higher-lying energy levels. In the case of the iron group ions it has long been known that most of their crystals are colored, and one of the early triumphs of crystal field theory was that it provided an explanation. Incident white light contains e.m. waves with frequencies that are able to induce transitions from occupied ground states to high-lying unoccupied levels at energies determined by the crystal field. An interesting observation is that these resonant waves are usually quite weakly attenuated, which implies that the transitions are weakly allowed. Most optical absorptions are, in comparison, much stronger, because they are usually allowed electric dipole transitions. That is, the transitions are induced, in first order of time-dependent perturbation theory, by the electric field of the radiation. In the case of a transition between two crystal field states that have come from the splitting of a "term," the initial and final states have the same parity and electric dipole transitions are forbidden. The strength of the absorption must be determined by something else; it is more likely to be a magnetic dipole transition, in which case it will be much weaker than an allowed electric dipole transition. (This was another success of crystal field theory, in that it provided an explanation for the energy differences in the optical range, and why the associated absorptions were so weak. The color of a fully concentrated magnetic crystal is enough to verify this. A transition due to an electric dipole transition would give the crystal a quite different appearance. More like coal!) The explanation of the colors of the rare earth salts is slightly different, for the crystal field splittings are usually only of the order of $100\,cm^{-1}$. The colors are therefore mainly due to transitions within the term structure, with fine structures due to spin-orbit and crystal field splittings. Their optical spectra tend to be much richer in lines than those found with iron group ions. They have provided an extensive field of study, particularly of the "so-called" crystal field splittings of excited levels (Judd, 1963). While it is also possible to study the thermally populated low-lying levels by optical spectroscopy, a more direct technique is to use inelastic neutron scattering. The neutrons lose measurable kinetic energies when they induce transitions between low-lying crystal-field-split levels (Furrer, 1977).

§ 5.2 Magnetic Susceptibilities

The spin-Hamiltonians usually omit the terms in the square of the magnetic field, which shift all levels by the same amount. Apart from a small temperature-independent contribution from these, the magnetic susceptibilities of assemblies of identical magnetic ions can be calculated directly from the spin-Hamiltonian. This involves using statistical mechanics to determine the free energy, F, the magnetic moment, M, and the susceptibility, χ, from the partition function, Z. The required formulae are

$$Z = \text{Trace}\left[\exp\left(\frac{-H}{kT}\right)\right], \quad F = -kT \log Z,$$

$$M = -\frac{\partial F}{\partial B}, \text{ and } \chi = \left(\frac{\partial M}{\partial B}\right)_{B\to 0}. \tag{5.1}$$

For a system of magnetic ions the H in the definition of Z is usually taken to be the spin-Hamiltonian of an isolated ion and F, M, and χ are multiplied by N, the number of ions under consideration. For a simple spin-Hamiltonian, such as $H = g\beta B S_z$, with $S = 1/2$, $Z = 2 \cosh(g\beta B/2kT)$, $M = (1/2)Ng\beta \tanh(g\beta B/2kT)$, and $\chi = N(g^2\beta^2/2kT)$.

In the early days of research on paramagnetism, particularly on crystals of the transition metal ions, Curie (1895) found a $1/T$ behavior in the susceptibility, a temperature dependence now known as Curie's law. It was derived theoretically by Langevin (1905) using an expression related to the present-day partition function. (Instead of using a discrete set the energies were regarded as forming a continuum, on the assumption that the magnetic moments would be found at all angles to the magnetic field.) However, as measurements were extended to lower temperatures it was found that the law needed modification, to the form $\chi = C/(T + \Delta)$, now known as the Curie-Weiss law.

The Langevin theory was extended by Weiss (1907). He returned to the expression for the magnetic moment in the limit when B is small, $M = CB/T$, realizing that the magnetic field acting on any particular moment ought to include the fields due to the magnetization of the other ions in the crystal. This led him to replace B by $B + \lambda M$, which immediately gave $\chi = C/(T - \lambda C)$, the Curie-Weiss form. The only problem was that Debye had already considered a related system, electric dipoles in a crystal with a cubic lattice, and had shown that for a spherical sample the internal electric field at any lattice site is zero. It was therefore difficult to see how the values of λC necessary to give Δ's equal to those observed could be due to internal magnetic fields. Weiss therefore suggested that there was some other source of internal magnetic field, adding that it was probably an *electrical* effect!

An examination of the modified expression for M showed that at low enough temperatures the expression has two solutions for M when B is set to zero. They are associated with different energies, and the one with the lower energy has a nonvanishing M. Weiss therefore concluded that, on lowering the temperature of a paramagnetic with $\chi = C/(T - \Delta)$, with Δ positive, there would be a phase change at $T_c = \Delta/k$ such that as the temperature was further lowered a spontaneous magnetization would rise from zero to a saturation value. At the time it was not possible to obtain low enough temperatures to test the prediction on the salts. There was, though, the example of a ferromagnetic, iron, and it was assumed that this was a system in which Δ is much larger than kT for room temperature, so that it is a paramagnetic which has already entered the ferromagnetic phase.

The theory was developed well before the advent of quantum concepts, and it is quite remarkable in the suggestion that the internal field is due to an electrical effect and in its prediction of a phase transition, for even today the general theory of phase transitions is a problematical area and one to which a good deal of attention is being paid. And yet the above demonstration seems so simple. The coming of the Bohr theory introduced the idea of discrete energy levels, so the Langevin theory was reworked, to give a form which was only slightly changed. With the Weiss substitution the expression for M becomes

$$M = 1/2 N g\beta \tanh\left[g\beta(B + \lambda M)/2kT\right].$$

(It is left to the reader to check that at low enough temperatures it too gives spontaneous magnetization when $B = 0$.) The origin of the internal field had to wait until the Coulomb interactions between electrons could be treated by quantum mechanics.

The susceptibilities for assemblies of free ions could be evaluated using the magnetic moments obtained from Zeeman spectroscopy, which gave, in present-day terms, the values of the effective spins and the g values to be inserted in the spin-Hamiltonians. For most of the iron group ions the agreement between the calculated and observed values was poor. With the coming of quantum theory and the introduction of electron spin, it became possible to check that the theoretical values of the spectroscopic g's, known as the Landé g factors, and the effective spins agreed with the spectroscopic observations, showing that the discrepancy was something to do with the ions being in a crystal. It was also soon realized that a much better fit to C could be obtained by neglecting the orbital contributions to the g factors.

In contrast, the C's for the relatively small number of rare earth crystals that had been studied agreed well with the g factors found from spectroscopy, suggesting that these ions were little affected by being in a crystal. More recent work has confirmed this, but it has also shown that the temperature dependences of the susceptibilities of many rare earth salts do not follow either the Curie or the Curie-Weiss law. There is no difficulty in understanding this, for the theory of ions in a crystal is in good shape. The difficulty is more in understanding why the early experiments gave misleading results. There are at least two possibilities—the measurements were seldom made on single crystals but on powders, and the susceptibilities of rare earth crystals often show pronounced anisotropy, which would be disguised in powders. There is also the possibility that the measurements were made on samples that contained unquantified amounts of other rare earths, for it is only comparatively recently that well-separated rare earths have become available. There is therefore some reason to view the early rare earth susceptibility results with reserve.

Surprisingly enough, a similar reservation seems necessary for the iron group results, though this time it is the values for the Δ's, rather than the C's, which cause concern. Looking at the early results (Weiss and Foëx, 1931), about half the samples showed positive values for Δ and half showed negative values. It is now established that the majority of iron group salts show values of the opposite sign to that predicted by the Weiss theory, implying that the internal field from the magnetization is in the opposite direction to the applied field.

A susceptibility of the form $C/(T + \Delta)$, where Δ is positive, might be expected to be well behaved down to $T = 0$. In fact, experiment has shown that this is not so, and that a crystal obeying this law at high temperatures will show a phase change at a temperature near Δ, into what is known as an antiferromagnetic phase.

§ 5.3 Exchange Interactions

The origin of an internal field of the strength necessary to account for the values of the Δ's remained a mystery for about twenty years, until Heisenberg (1926) and Dirac (1926) independently published their work on a quantum mechanical problem which can be regarded as an example of first-order degenerate perturbation theory. They assumed that N electrons had been placed, one by one, into N orthonormal one-electron orbital states, to give a P_0 which spanned a 2^N-fold degeneracy arising from the different Slater determinantal states that arise on allowing each spin to have two possible orientations. The problem was to determine the splittings of P_0 that would arise on introducing a Coulomb interaction between the electrons as a perturbation. Finding solutions to the problem with a general value of N still presents difficulty, though for just two electrons it is both soluble and instructive. Because the Hamilton is rotationally invariant in spin space the eigenstates can be classified according to their total spins. The two spins of $1/2$ give four states which can be arranged to form an $S = 1$ triplet, leaving just one other possibility, an $S = 0$ singlet. The three $S = 1$ states are

$$\{a^+, b^+\}, \quad \frac{1}{\sqrt{2}}[\{a^+, b^-\} + \{a^-, b^+\}], \quad \{a^-, b^-\} \qquad (5.2)$$

and the $S = 0$ state is

$$\frac{1}{\sqrt{2}}[\{a^+, b^-\} - \{a^-, b^+\}], \tag{5.3}$$

where a and b denote the orbital states, the $+$ and $-$ superscripts denote the spin orientations, and the $\{.,.\}$ that the states are Slater determinants. The energy of each $S = 1$ state is then equal to

$$\langle a, b|(e^2/r_{12})|a, b\rangle - \langle a, b|(e^2/r_{12})|b, a\rangle$$

and the energy of $S = 0$ is

$$\langle a, b|(e^2/r_{12})|a, b\rangle.$$

The triplet is below the singlet because the energy difference is given by an exchange-type matrix element, $\langle a, b, |(e^2/r_{12})|b, a\rangle$, which is known to be positive. Also, with the P_0 manifold being isomorphic with a manifold of two spins of $1/2$, it is possible to derive a spin-Hamiltonian to describe the energy level pattern. Using the result that

$$2\mathbf{s}_1 \cdot \mathbf{s}_2 = (\mathbf{s}_1 + \mathbf{s}_2) \cdot (\mathbf{s}_1 + \mathbf{s}_2) - \mathbf{s}_1 \cdot \mathbf{s}_1 - \mathbf{s}_2 \cdot \mathbf{s}_2,$$

it readily follows that the operator

$$K_{ab} - (3/4)J_{ab} - J_{ab}\mathbf{s}_1 \cdot \mathbf{s}_2,$$

where $K_{ab} = \langle a, b|(e^2/r_{12})|a, b\rangle$ and $J_{ab} = \langle a, b|(e^2/r_{12})|b, a\rangle$ has the same level pattern as the original problem.

The investigation of the N-electron problem did not yield the eigenvalues and eigenstates, but it showed that it was possible to find an operator, which can be regarded as a spin-Hamiltonian, that would have the same energy level pattern and the same degeneracies. Its structure was similar to that derived for the two-electron case, except for summations over all possible K_{ab} and J_{ab}, with each of the latter multiplying a different $\mathbf{s}_1 \cdot \mathbf{s}_2$ operator pair, which therefore appeared as $\mathbf{s}_a \cdot \mathbf{s}_b$. If each orbital state is then regarded as attached to a different nucleus, $\mathbf{s}_a$ can be regarded as the effective spin operator at the site labeled a.

The similarity of the model with that of a lattice of magnetic ions, each having a single electron in a singly degenerate crystal field ground state, was quickly recognized. A reasonable assumption then seemed to be that the exchange interactions, the J_{ab}'s, would only be appreciable between nearest-neighbor ions and, for a crystal in which the magnetic ions are cubically arranged or in an equally spaced linear chain, they would be identical for all nearest-neighbor pairs. The problem of finding the eigenvalues and eigenstates of what is now known as the Heisenberg exchange interaction,

$$\mathcal{H} = J \sum_{\langle i,j \rangle} \mathbf{s}_i \cdot \mathbf{s}_j, \tag{5.4}$$

where the summation is taken over all nearest-neighbor pairs and each spin is regarded as having spin $1/2$, is an example of a problem that has arisen from what appears to be a simple physical situation. It has now been taken over as a challenging problem in mathematics (Mattis, 1985; Caspers, 1989; and Oitmaa et al., 1994, as a source of more recent references). The distinction is worth emphasizing for it is not at all certain that there is any physical system for which it is an exact model, which illustrates a not uncommon practice, that of finding approximate eigenvalues and eigenfunctions of some effective Hamiltonian and using an apparent agreement with experiment to justify the approximation and the use of the effective Hamiltonian.

Returning to the Curie-Weiss Δ, it could be inferred that the apparent ferromagnetic values arose from a combination of the antisymmetric requirement on the wave function and the two-electron nature of the Coulomb interaction. They could evidently produce spin-Hamiltonian-like interactions of ferromagnetic sign between spins on neighboring ions. It was, of course, necessary to show that such interactions actually give the required Curie-Weiss form for the susceptibility (see Van Vleck, 1937a; Hudson, 1972, §6.2). The same analysis, with dipole-dipole interactions between the spins of the form

$$\frac{g^2\beta^2}{r_{ij}^3}\left[\mathbf{s}_i \cdot \mathbf{s}_j - 3\frac{(\mathbf{s}_i \cdot \mathbf{r}_{ij})(\mathbf{s}_j \cdot \mathbf{r}_{ij})}{r_{ij}^2}\right],\tag{5.5}$$

which is the quantum form of the classical interaction between two moments a distance r_{ij} apart, confirms that they make no contribution to Δ for a cubic lattice of spins.

Neither the exchange nor the dipolar interaction emerges directly from crystal field theory, so any such extensions of the spin-Hamiltonian concept as applied to magnetic ions must be regarded as going beyond the usual treatment of crystal fields. In fact, it is not at all obvious how crystal field theory can be so extended. This has not prevented various authors from suggesting that with two adjacent magnetic ions the Hamiltonian should be extended so as to contain the Coulomb interactions between the electrons on the different ions, and that these might be dropped in a first approximation. Each ion would then be described by its own spin-Hamiltonian. The neglected interactions, when they are incorporated using perturbation theory, should then lead to an exchangelike coupling between the two effective spins. So far, though, no satisfactory way of doing this has emerged, one reason being that generating a separate spin-Hamiltonian for each ion would appear to distinguish between the electrons on the two ions and lead to an unusual feature in perturbation theory, that the introduction of the perturbation would increase the symmetry. It is not clear how this would show up in the construction of the states of the "perturbed" system from linear combination of the states of the unperturbed system.

As more crystals were studied it became apparent that in general the Δ's had the wrong sign for ferromagnetism and that below the antiferromagnetic co-operative phase change at T_c adjacent moments are antiparallel. Indeed, many crystals have magnetic lattices that can be regarded as made up of interpenetrating A and B sublattices, each being a ferromagnetically aligned lattice, with the moments on the A sublattice being oppositely aligned to those on B. The Weiss theory, extended to such a system (Bitter, 1937), showed that above T_c the susceptibility would follow a Curie-Weiss law but below T_c it would depend on the direction of the applied field relative to the alignment direction. For a parallel arrangement it would uniformly decrease to zero, and for the transverse arrangement it would remain constant. There would therefore be no striking change in the magnetic properties, such as would be expected with a ferromagnetic interaction (Bizette et al., 1938). For some years the concept of antiferromagnetism seems to have been rather reluctantly accepted. The position change drastically in the early 1950s when two new techniques, EPR and neutron scattering, came into use.

An EPR study of copper acetate (Bleaney and Bowers, 1952a,b) showed that the Cu^{2+} ions, which from previous EPR in other hosts were expected to have effective spins of $1/2$ and slightly anisotropic g values near 2, gave resonances that could only be interpreted as coming from two closely spaced Cu^{2+} ions coupled by an exchange interaction of antiferromagnetic character and of magnitude $310\,cm^{-1}$. The ground state of the pair, having $S = 0$, would show no resonance, but at high enough temperatures the $S = 1$ level would be thermally populated and show resonances. From the temperature dependence of the signals Bleaney and Bowers were able to determine the magnitude of the exchange interaction, and from the details of the spectrum they were able to infer that while it was dominantly of the Heisenberg $s_1 \cdot s_2$ type there was evidence that a small amount of what is known as anisotropic exchange was also present.

The concept of anisotropic exchange had been introduced by Van Vleck (1937b), because the Heisenberg interaction has the slightly unfortunate property that it is isotropic in spin space. So if, below a T_c, it does cause an alignment of spins, it provides no indication of the alignment direction. The same point had concerned Weiss in applying his ideas to the ferromagnetism of iron, for it was known that freshly prepared iron has no magnetic moment. This had led him to introduce the idea of domains, regions in which the moments are aligned in specific directions relative to the crystallographic axes. A polycrystalline sample would initially have no overall moment, but an external field could change all this by making the various domains take up new orientations. It was, however, evident that some other unidentified interaction was present, to make the moments align preferentially in crystallographic directions. (Magnetic dipole interaction between moments, of the form given in (5.5), was a possibility, except that for cubic lattices they seemed too small.) With the increased understanding of ions in crystal fields, another possibility was that magnetic anisotropy could be associated with the zero-field splittings, except that they vanish for $S = 1/2$, which would seem to rule out this explanation for copper acetate. For ions with zero-field splittings there is no doubt that the ions may be anisotropic, depending on the environment, but if so it is just as apparent in the magnetic moments as in the zero-field splittings, both of which arise from spin-orbit interactions in second order of perturbation theory. Van Vleck therefore suggested that there could be exchangelike interactions between adjacent ions due to spin-orbit interactions, and that they would have the same general form as the dipole-dipole interaction of (5.5), except that the factor of $(g^2\beta^2/r_{ij}^3)$ would need to be replaced by a short-range function of r_{ij}. (Both the dipolar interaction and this new one are invariant under rotations of axes. The positional parts and the spin parts separately transform as D_2 and the overall interaction is the D_0 component of $D_2 \times D_2$.) It was later suggested (Stevens, 1953) that there could also be, in some cases, another invariant interaction, one that is again of short range but antisymmetric in the spins of the two sites, with the form $(s_1 \wedge s_2) \cdot \mathbf{a}$, where $\mathbf{a}$ is a vector in spatial variables. (This interaction is now known as the Moriya-Dzialoshinski interaction.)

At the time when Van Vleck introduced anisotropic exchange, the concept of a spin-Hamiltonian had not come into general use, but by the 1950s the position had changed, and isotropic, anisotropic, and antisymmetric exchange interactions of the above forms were being introduced into spin-Hamiltonians of lattices of magnetic ions, even though there was no reason to suppose that the interaction between two ions, both of which were likely to have had their energy levels and states substantially perturbed by being in a pair, would be invariant when expressed in spin operators. The best that can be said is that if the exchange interactions are linear in the two effective spins then the above forms cover all possibilities.

Another development arose from the work of Shull et al. (1951), who showed, using neutron scattering, that the antiferromagnetic exchange interactions between the Mn^{2+} ions in MnO, ions that were expected to have $L = 0$ and $S = 5/2$, had stronger interactions when they were separated by O^{2-} ions than when they were actually closer together but with no intervening O^{2-} ions. The T_c for the transition, now known as a Néel temperature, was 122 K, and the assumption was that this was determined by an isotropic antiferromagnetic exchange interaction of cosine form between spins of $5/2$. So the original Heisenberg-Dirac form was given a further extension. The sign had already been reversed; now the spins of $1/2$ were extended to $5/2$.

The increasing knowledge about "exchange," as obtained by experiment, stimulated an interest in providing a theoretical description that went beyond the Heisenberg-Dirac theory and beyond crystal field theory. A variety of "exchange processes" were postulated and expressions such as "direct," "indirect," and "super" exchange began to appear in the literature, expressions that no doubt had a clear meaning to those who introduced them but which were not necessarily clear to every reader. The basic difficulty appears to have arisen because in using perturbation theory there is a need to choose unperturbed states and these do not have to be mutually orthogonal, though if they are not the perturbation formulae are much more complicated than if they are. Then it became a common practice, particularly in going beyond first order, to give specific names to processes thought to be important (e.g., a name is given to a process in which an excited state has been obtained by moving an electron from some occupied orbital to some unoccupied orbital and then back again), which is all right provided the process is fully described. But when there is some uncertainty about the perturbation method, the nature of the states involved, and the operator inducing the transition, the description becomes ambiguous. No attempt will therefore be made to define the meanings of the various types of exchange listed above. The various possible processes were codified and put on a firmer foundation in a series of papers by Anderson and brought together in a review (Anderson, 1963).

§ 5.4 Delocalized Orbitals

It is now of interest to return to the Heisenberg-Dirac theory, for it has a feature that has so far not been emphasized, that it is based on orbitals which are mutually orthogonal. Van Vleck (1935) applied crystal field theory to a few compounds which were described as covalent, the cyanides of the $(3d)$ ions, knowing that only some of the iron group elements form them, presumably because with the others they would not be chemically stable. Unlike the case of the ionic crystals he was led to assume that, in the stable complexes, the crystal fields are so strong that the term structures are disrupted. So instead of Fe^{3+}, in a crystal, having five electrons in the $(3d)$ shell, which would result in the lowest term having $S = 5/2$, the octahedral crystal field has already split each $(3d)$ state into a low-lying $\{xy, yz, zx\}$ triplet and an upper $\{3z^2 - r^2, x^2 - y^2\}$, doublet with the five electrons having gone into the triplet to give a spin of $1/2$. With six nominally $(3d)$ electrons the subshell would be filled. This model accounted for the observed moments and provided an explanation of why there were no cyanide complexes for ions that could be expected to have more than six d electrons. To a reasonable degree the model also explained the EPR results on the iron group cyanides, obtained much later, though accounting for the magnitudes of the spin-orbit interactions, inferred from parameters in the spin-Hamiltonians, raised questions. In due course, EPR experiments were carried out on some similar complexes containing ions of the $(5d)$ series, and in $IrCl_6$ Owen found a hyperfine structure with features that had not been seen before. These were interpreted (Owen and Stevens, 1953) as arising because the orbitals of the electrons, which it had previously been assumed would be like the (5d) orbitals of a free ion, were distorted in the complex and contained admixtures of states of the chlorine ions. There would therefore be unpaired electron spins on the ligands, which, having chlorine nuclei with magnetic moments, gave the additional hyperfine structure. So the explanation of the properties of the $IrCl_6$ complex went beyond the strong-field crystal field model and needed a description in terms of delocalized orbitals, similar to those being used for molecules. [It was subsequently realized that in most of the previous EPR work on (3d) ions the neighboring ions, the ligands, were O^{2-} ions, the overwhelming majority of which have nonmagnetic nuclei. When (3d) ions with ligands that had nuclear magnetic moments were substituted, transferred hyperfine structures were again found, showing that the magnetic electrons were not confined to the central ion.] These discoveries made it unlikely that the so-called (3d) orbitals were as simple in their angular properties as the (3d) orbitals of the free ions, and that the orbitals on adjacent magnetic ions in a crystal would be orthogonal to one another or to those of the central ion. For the isolated ions it was enough to add extra terms, very much like those already present, to the spin-Hamiltonian, to describe the hyperfine couplings to the ligand nuclei. The EPR results, being on isolated complexes, threw little light on the pair interactions, apart from indicating that some modifications might be needed because the Heisenberg-Dirac theory assumes that the electrons are in orthogonal orbitals. The few changes that did result were in the isolated complex theory, and led to the idea of reductions in the magnetic moments of orbitals on delocalization and a theory of the hyperfine interactions of ligand nuclei. It can now be seen that these were minor compared with the changes which came later and which have transformed the way of looking at magnetic ions in crystals. These will be described in detail in due course.

As an introduction it is of interest to consider a pair of ions in which the orbitals on one ion are fairly obviously not orthogonal to the orbitals on the other. This is provided by the hydrogen molecule. On the simplest picture it consists of two closely spaced hydrogen atoms, A and B. Atom A might be expected to have an electron with an unpaired spin in a 1s orbital centered on nucleus A and atom B might be expected to have the other electron in a 1s orbital centered on B, a description that makes the molecule appear similar to two closely spaced magnetic ions. It is very convenient in many pieces of theory either to ignore the fact that electron orbitals are not orthogonal or to assume they are. It is also obvious that the present two functions cannot be orthogonal, for both have the same sign everywhere. To progress a choice seems necessary, either to develop a theory using nonorthogonal orbitals or to combine the nonorthogonal functions so that they give orthogonal functions. The latter procedure will be followed.

If $|A\rangle$ denotes a normalized (1s)-like state at one nucleus and $|B\rangle$ denotes the corresponding state at the other nucleus, $|A\rangle$ and $|B\rangle$ will be mirror images and the combinations $[|A\rangle + |B\rangle]$ and $[|A\rangle - |B\rangle]$ will be mutually orthogonal. They become orthonormal if the first is divided by $[2(1 + S)]^{1/2}$ and the second by $[2(1 - S)]^{1/2}$, where S is the overlap, $\langle A|B\rangle$. (Combinations such as $[|A\rangle + |B\rangle]$ etc., which involve orbitals at more than one site, are known as molecular orbitals.) With two electrons it is therefore possible to construct six different Slater determinants, three forming a degenerate $S = 1$ triplet and three with $S = 0$. So the ground state will be either a triplet with $S = 1$ or a singlet with $S = 0$. Experiment shows that it is a state with $S = 0$. (It is usually thought of as a state that approximates to having both electrons in the symmetric combination $[|A\rangle + |B\rangle]$, with opposed spins.)

The molecule can be looked at in another way, for with two orthonormal orbitals an infinite number of other orthonormal pairs can be constructed, since

$$[\cos\theta|U\rangle + \sin\theta|V\rangle] \quad \text{and} \quad [\sin\theta|U\rangle - \cos\theta|V\rangle]$$

are mutually orthogonal for all choices of θ if $|U\rangle$ and $|V\rangle$ are orthonormal. In the present example, taking $|U\rangle$ to be one of the orthonormal combinations of $|A\rangle$ and $|B\rangle$ and $|V\rangle$ the other, there is a choice for θ which minimizes the amount of $|B\rangle$ in the first member of the pair and maximizes it in the second, to give a pair of states that are orthonormal and localized as much as possible to the respective nuclei. The Heisenberg-Dirac theory would then predict that with one electron in each of these localized orbitals the $S = 1$ level will be below the $S = 0$ level, which is not what is observed. What the treatment has overlooked is that there are two more $S = 0$ states, which can be obtained by putting both electrons, with paired spins, into either of the localized orbitals, and that in the full treatment these states need to be taken into account, particularly as in second order of perturbation they will give a lowering of the lowest $S = 0$ state. The experimental $S = 0$ ground state can therefore be understood as due to the presence of admixtures of these extra states into the $S = 0$ unperturbed ground state, so putting the lowest $S = 0$ state below the $S = 1$ triplet. Since it does not matter which orthonormal basis is used in a full perturbative treatment, this $S = 0$ will be identical with that obtained with the orthonormal molecular orbitals. [Unlike the Heisenberg model the detailed calculation includes the kinetic energy of the electrons as well as their Coulomb energy; see Slater (1968), chap. 21.]

From the existence of many covalently bonded molecules and from the paramagnetism of many metals, it can be concluded that a good physical description can usually be obtained by assuming that electrons occupy orthonormal orbitals in pairs, so that spin pairing is the common arrangement. Parallel spin arrangements are almost entirely confined to isolated ions with partially filled shells, to a small number of ferromagnetic metals and alloys, and to the magnetic crystals that contain iron and rare earth group ions. And even in the magnetic crystals

Table 5.a. The Character Table of the C_3 Group, Defined by T

C_3	E	C_3	C_3^2
A	1	1	1
E_1	1	ω	ω^2
E_2	1	ω^2	ω

the usual arrangement has, as far as possible, an antiparallel alignment of the spins on adjacent magnetic ions.

There are a few exceptions, O_2 being an interesting example of a gas with an even number of electrons and $S = 1$. Using a molecular orbital description, the highest occupied orbital level is degenerate for symmetry reasons (the molecule has an axis of rotation and inversion symmetry). With most of the electrons being paired off in low-lying orbitals, two remain to be placed in an energy level that is an orbital doublet. The $S = 1$ arises because the lowest energy occurs when the two electrons are in different orbital states of the degenerate pair with parallel spin alignments. This arrangement is so similar to that found in the free ions that it can be supposed that electrons distributed over orbitals that are degenerate for symmetry reasons generally occupy them one by one, with as many spins as possible having parallel alignments.

Having introduced the concept of molecular orbitals in connection with the H_2 molecule, it is now of interest to consider how mutually orthogonal orbitals might be constructed for a system of three hydrogen atoms arranged to form an equilateral triangle. Denoting the 1s orbitals by $|A\rangle$, $|B\rangle$, and $|C\rangle$ the three combinations $|A\rangle + x|B\rangle + x^2|C\rangle$, where x ranges over the three cube roots of unity, are mutually orthogonal. (With four hydrogens arranged to form a square, the combinations $|A\rangle + x|B\rangle + x^2|C\rangle + x^3|D\rangle$, with x ranging over the four fourth roots of unity and $|D\rangle$ denoting the orbital of the fourth hydrogen, are similarly orthogonal.) There are various ways of proving the orthogonality; here it will be done using group theory.

For either example there is an operator, T, which moves each hydrogen into the position occupied by a neighbor. With three hydrogens it readily follows that after T has been used three times the initial position is restored. T can therefore be regarded as defining a rotation group that consists of just three commuting elements E, T, and T^2. Each operator therefore defines a class consisting of one element, and the character table, which is particularly simple, can be written down by inspection or by using the general properties of group characters. [See table 5.a where $\omega = \exp(2\pi i/3)$.]

If T takes $|A\rangle$ into $|B\rangle$, etc., it takes $(|A\rangle + x|B\rangle + x^2|C\rangle)$ into $(|B\rangle + x|C\rangle + x^2|A\rangle)$, which is equal to $x^2(|A\rangle + x|B\rangle + x^2|C\rangle)$. Thus the choice $x = 1$ gives a basis element for A, the choice $x = \omega$ gives a basis for E_2, and the choice

$x = \omega^2$ gives a basis for E_1. Each choice for x gives a basis element for a different i.r. The orthogonality follows from the decomposition of products of the i.r.'s, because a matrix element of the form $\langle a|b \rangle$ is zero unless the i.r. of $\langle a|$ is the adjoint of the i.r of $|b\rangle$. (In the present example, E_2 is the adjoint of E_1 and it is the product representation $E_1 \times E_2$ which gives A. If $|a\rangle$ transforms as E_1 and $|b\rangle$ as E_2, then $\langle a|b \rangle$ transforms as $E_2 \times E_2$, which is E_1. $\langle a|b \rangle$ is therefore zero.)

With four hydrogens it will be T^4 that gives the unit operator, so there will be four classes and four one-dimensional i.r.'s. In the part of the table that gives the characters the first column, E, will have a sequence of 1's, the second column, the class of T, will have the four roots of unity in sequence, in the next, the class of T^2 they will be squared, and so on. The procedure, which can clearly be extended to an arrangement of N equally spaced hydrogen atoms in a circle, when it becomes more like a translational group, will give a group which has one-dimensional i.r.'s, one for each of the Nth roots of unity. The character table will be an extension of that for $N = 3$ and will be so simple that there is no need to display it.

The three-site example and its extension to a circle of hydrogen atoms has been chosen to pave the way for describing a lattice of magnetic ions, for they illustrate a feature that arises when identical orbitals, on different sites, are not orthogonal: that each of the obvious orthogonal orbitals that can be constructed from them spreads over all the sites. They are much more like molecular orbitals, though the common description in solid state physics is that they are delocalized, in contrast with the orbitals used in crystal field theory, which are described as localized.

§ 5.5 Magnetic Conductors

The commonest examples of ferromagnetism occur with iron, cobalt, and nickel and their alloys. They are all electrical conductors, whereas the commonest examples of antiferromagnetism are found with the iron group salts, which are insulators. Another difference is that if the magnetizations per site are compared with those for the antiferromagnets, the latter correspond closely to having a whole number of aligned spins at each magnetic site, whereas those in the ferromagnets, nickel in particular, show moments which suggest that there is a fraction of an unpaired spin at each site. [Measurements have shown that the changes in angular momentum that occur on magnetization are due to electron spin (see Gallison, 1987).] Later, using ferromagnetic resonance, it was found that in nickel there is a small contribution from an orbital moment (Griffiths, 1951). The 0.6 value usually given for the number of unpaired electrons at each site in nickel should therefore be changed to 0.5. Iron and cobalt have values nearer to integers (Bates, 1961; see table XXV, p. 317).

Fractional numbers of unpaired spins at identical sites presented something of a problem initially, for it was assumed that electrons occupy localized orbitals. It was not long, though, before a way out of the difficulty was found. The Coulomb interactions between the electrons were replaced by an effective potential, which, though it had the periodicity of the lattice, was assumed to be almost uniform. This established a link between the description of an electron in a lattice and the standard quantum mechanical problem of the motion of a particle in a box. For the box problem the energy levels form what is almost a continuum above a lowest value. When the motion in a periodic potential was examined, using what are known as periodic boundary conditions, it was found that the continuum of energy levels of the uniform potential splits into bands of levels. The ground electronic state then has the electrons, two by two and spin paired, occupying the band states up to a certain energy, above which the states are empty. There is then a continuum of energy levels above the ground energy level as electrons are moved from states below the highest occupied one-electron level to states above it.

This model readily explained the origin of the conductivity and the temperature-independent diamagnetism of most metals. Somewhat more tentatively, it promised an explanation of the ferromagnetism of iron, cobalt, and nickel, for the assumption of a periodic potential is conceptually similar to the central potential in the free ions, where it was well established that on improving the approximation spin alignments occur in partially filled shells. If the same holds for electrons in partially filled bands it would seem that a total spin, S, could have a magnitude which is unrelated to that of an integral number of aligned spins at each site. Giving a rigorous demonstration has proved difficult, though it is almost certain that where ferromagnetism does occur it is due to the Coulomb interactions between electrons.

§ 5.6 Conflicting Theories

By the early 1950s the use of band theory to describe periodic structures was very much in the ascendance, so much so that to its exponents it seemed only a matter of time before it would explain ferromagnetism and, probably, all phenomena associated with periodic lattices. Yet it was at this same time that crystal field theory was having its greatest success. So two apparently incompatible schools of thought were being developed, one firmly wedded to delocalized orbitals as the way in which to deal with all periodic systems, and the other, with its localized concepts, having established, experimentally, the validity of spin-Hamiltonians for describing isolated magnetic ions. (And seemingly well on the way to incorporating the interactions between the ions within the framework of the same formalism.)

Looking back it seems not unreasonable to claim that as a result of this divergence the study of magnetism was split into two for something like twenty years, a position which could be maintained because neither school could explain the observations of the other. Nor could either cope with what is the oldest problem in magnetism, the explanation of the ferromagnetism of iron, cobalt, and nickel.

The position began to change in the late 1950s when the magnetic properties of the rare earth metals became of interest. Both approaches seemed to be needed, band theory to explain that the conduction properties were due to mobile delocalized outer-shell electrons and crystal field theory to explain that the magnetic properties were due to inner-shell immobile and localized ($4f$) electrons. Each school was thus being pressed into becoming more aware of the other, though with such a fundamental divergence between them it was not to be expected that conciliation would occur without major changes somewhere.

The emphasis in the previous chapter has been on describing the development of crystal field theory. It now seems only proper to devote some attention to band theory, if only so that the two approaches can be compared. Having done so, it becomes possible to see that a common ground can, perhaps, be found. Most of the subsequent chapters will therefore be devoted to examining this in more detail.

§ 5.7 Band Theory

The point has already been made that for solid state purposes it can be assumed that all systems have basically the same Hamiltonian, with the only variations coming from changes in the nuclear charges and the number of electrons. So its properties should not depend on whether it is treated by band theory or by some modified version of crystal field theory. Band theory arises from an approximation which separates the unperturbed many-electron Hamiltonian into a sum of one-electron Hamiltonians, each of which has the form

$$H = \frac{\mathbf{p}^2}{2m} + V(x, y, z), \tag{5.6}$$

where $V(x, y, z)$ is a potential energy which has the periodicity of the lattice and which extends to infinity in all directions. The unperturbed Hamiltonian then has a number of symmetry properties, for not only is it invariant under the point group operations that take any chosen unit cell into itself, but it is also invariant under displacements that take one unit cell into another. To avoid having to deal with an infinite system, it is usual to assume that the lattice folds back on itself, so that after a large but finite number of displacements each unit cell comes back to itself (the analogue in three dimensions, where it is actually not possible, except mathematically, of the model in which N hydrogen atoms are equally spaced round a circle).

Three translation operators can be defined, T_a, T_b, and T_c, where T_a, for example, is a displacement through one unit cell in the direction of the side a of the unit cell. T_b and T_c are then the operators for the corresponding displacements in the directions of the sides b and c of the unit cell. With periodic boundary conditions, each T when raised to the appropriate high power equals E, the unit operator in the translation group defined by T_a, T_b, and T_c. As these operators commute, the irreducible representations of the translational group can be characterized by the appropriate Nth roots of unity. If the lattice repeats itself after N_1 steps in the a direction, N_2 in the b direction, and N_3 in the c direction, the set of integers (p, q, r) can be used to denote a particular i.r., one in which the characters of T_a, T_b, and T_c are respectively $\exp(2\pi i p/N_1)$, $\exp(2\pi i q/N_2)$, and $\exp(2\pi i r/N_3)$. [The values of p, q, and r have to be restricted to what is known as the reduced Brillouin zone, for increasing p by an integer multiple of N_1 does not alter the value of $\exp(2\pi i p/N_1)$. So for the i.r.'s to be unique one of a range of possible restrictions is chosen.] The eigenfunctions of the one-electron Hamiltonian, (5.6), can be used as basis elements for such a set of i.r.'s. It follows that the eigenfunction that is a basis element for the (p, q, r) i.r. has the property

$$\psi(x+la,\, y+mb,\, z+nc) = \exp\left[2\pi i\left(\frac{lp}{N_1} + \frac{mq}{N_2} + \frac{nr}{N_3}\right)\right]\psi(x, y, z), \quad (5.7)$$

where a, b, and c are the lengths of the sides of the unit cell and l, m, and n are integers. On setting $l = N_1$, $m = N_2$, and $n = N_3$, $\psi(x+la, y+mb, z+nc)$ becomes $\psi(x, y, z)$, showing that the periodic boundary conditions are satisfied. (The cell with sides $N_1 a$, $N_2 b$, and $N_3 c$ will be referred to as the extended cell.) It may be noticed that for most values of p, q, and r, $\psi(x, y, z)$ is *not* a periodic function of l, m, or n, though its square modulus is. The triads (p, q, r) can be represented as points in a three-dimensional lattice in which the unit cells have sides of length unity. However, as p usually appears in the combination $2\pi i p/N_1$, and similarly for q and r, it is more convenient to use (k_x, k_y, k_z) instead, where $k_x = 2\pi p/N_1$, $k_y = 2\pi q/N_2$, and $k_z = 2\pi r/N_3$, to define what is known as the reciprocal lattice. The points (k_x, k_y, k_z) are much closer together since the whole range of a possible k value is only 2π. If the size of the extended cell is increased the volume of k space remains unaltered, so the density of points in it tends to a continuum. As will be seen, the eigenvalues of (5.6) vary with $\mathbf{k} = (k_x, k_y, k_z)$. It follows that as the size of the extended cell is increased the separation between adjacent $\mathbf{k}$'s tends to zero. It then becomes possible to regard both $\mathbf{k}$ and its associated eigenvalues as continuous functions.

For a Bravais lattice with a cubic unit cell there are many symmetry operations, rotations, reflections, etc., which leave the extended cell invariant, provided that $N_1 = N_2 = N_3$ (which usually leads to no loss in generality). It might then be expected that if one of the eigenfunctions is examined in the vicinity of some nucleus it will have the same symmetry properties under the point group operations as would be found for one of the eigenfunctions of a hydrogen atom in a cubic (or octahedral) crystalline potential. If so, it could then be put into correspondence, locally, with one of the basis functions that generates an i.r. of the cubic group. In fact this is incorrect, for the wave functions change too much from cell to cell, so even if it is true for one cell it will not generally be true for all the cells.

This is best seen with the "tight-binding model." Suppose that $\phi(\mathbf{R}_0)$ is an atomic-like wave function defined with respect to an origin in a particular unit cell, the cell at $\mathbf{R}_0$. Then a family of functions, $\phi(\mathbf{k})$, known as Bloch functions, can be defined by

$$\phi(\mathbf{k}) = \frac{1}{S(\mathbf{k})} \sum_{\mathbf{R}_n} \phi(\mathbf{R}_n) \exp(i\mathbf{k} \cdot \mathbf{R}_n), \tag{5.8}$$

where $\phi(\mathbf{R}_n)$ is the same function as $\phi(\mathbf{R}_0)$, except that it is centered in the cell at $\mathbf{R}_n$, $S(\mathbf{k})$ is a normalizing factor, and $\mathbf{k} = (k_x, k_y, k_z)$. If now the cells are relabeled, so that each $\mathbf{R}_n$ becomes $\mathbf{R}_n + \mathbf{a}$, where $\mathbf{a}$ is the vector describing one of the edges of a unit cell, $\psi(\mathbf{k})$ becomes $\exp(i\mathbf{k} \cdot \mathbf{a})\psi(\mathbf{k})$, showing that $\psi(\mathbf{k})$ can be used as a basis element for the i.r. defined by $\mathbf{k}$. It follows that as $\mathbf{k}$ ranges over the reduced Brillouin zone (5.8) generates an orthonormal set of functions. It also seems probable that if $\phi(\mathbf{R}_0)$ is chosen to be a suitable hydrogenic ($1s$) function the corresponding $\phi(\mathbf{k})$ functions, which are analogous to those introduced in considering the chain of hydrogen atoms, might be a good approximation to an eigenstate of (5.6). (This is the assumption of the tight-binding model.)

A better approximation might be obtained by beginning with a set of hydrogenic functions defined in the $\mathbf{R}_0$ cell and constructing a $\phi(\mathbf{k})$ set from each of them. Those with different $\mathbf{k}$'s would be orthogonal. Those with the same $\mathbf{k}$ would not be, in general. So a further step would be to use an orthogonalizing process, for each $\mathbf{k}$, to generate an orthonormal set. It then follows that (5.6), being invariant under the translation group, will have zero matrix elements between functions with different $\mathbf{k}$'s. So the choice of the tight-binding functions has partially separated (5.6) into sub-Hamiltonians of different $\mathbf{k}$'s. In many cases the second step is not carried out, and it is simply assumed that the $\phi(\mathbf{k})$'s obtained with the different hydrogenic choices give expectation values of (5.6) that are good enough approximations to eigenvalues. They will clearly depend on the choice of $\mathbf{k}$, but as this belongs to a distribution which approaches a continuum it can be expected, beginning with a particular hydrogenic function, $\phi(\mathbf{R}_0)$, that the expectation values will have a distribution almost indistinguishable from a continuum. For each hydrogenic function there will therefore be a band of energy levels, though because of the restrictions on $\mathbf{k}$ the number of states in any band will be equal to the number of unit cells in the extended cell. Taking the step of orthogonalizing the ϕ's of a given $\mathbf{k}$ can be expected to change all the energy levels, but it is unlikely to remove the picture of a set of bands, one associated with each hydrogenic function.

The hydrogenic functions have radial wave functions which go to zero at infinity, so such a function centered in the R_0 unit cell will extend outside the cell. In constructing a $\phi(k)$ the same functions, but centered in other unit cells, are added with k-dependent phase factors. The tails of these will have values within the R_0 unit cell. It is the tails which prevent most of the $\phi(k)$'s, when examined within any particular cell, having a hydrogenic form.

The ϕ's with $k = 0$ are particularly interesting because in these the hydrogenic functions are simply added together, so a $1s$ function in one cell produces a $(1s)$ function in every cell. For a cubic lattice any operation of the point group symmetry then leaves the ϕ in every cell unaltered. Similarly, if the hydrogenic functions in the central cell are chosen to transform as elements of an i.r. of the cubic group, then the set of associated ϕ's that have $k = 0$ also forms a basis set for the same i.r. This reasoning, which can be extended to any shape of unit cell, leads to a way of labeling bands. All that is necessary is to examine each ϕ which has $k = 0$. Within any cell it can be used to generate one or more irreducible representations of the local point group. These can be used to label the bands that have these $k = 0$ states in them, though it does not follow that the states with $k \neq 0$ have the same symmetry properties or the same degeneracies. (At a general k they usually have no symmetry properties.)

For a cubic lattice the hydrogenic p functions with a given n give rise to three bands which all have the same energy at $k = 0$. They can therefore be labeled as p bands or by T_1. For the d functions of a given n, in a cubic lattice, the hydrogenic functions span two i.r.'s, E and T_2. There will be five associated bands, three of which will have the same energy at $k = 0$ and can be denoted by T_2, and two more, which have the same energy as one another at $k = 0$, but an energy different from that of the T_2 band, which can be labeled by E. (In many textbooks these distinctions are not made and the bands are simply labelled as s-, p-, and d-like.)

For non-Bravais lattices the same basic procedure can be followed, though now the Bloch functions are constructed from the one-electron states of each ion in the unit cell. For each k value there will be more Bloch functions, so the one-electron Hamiltonian of (5.6) generally has more band states of the same k in any energy range. The resulting band structures are usually more complicated, as is their labeling at $k = 0$.

An approximate ground state of the many-electron system, the analogue of a configuration for an ion, is obtained by placing the electrons into the band states, two by two with their spins paired, in increasing order of energy. The construction of each Bloch function shows that the number of states in any band is equal to the number of different k values, which in turn is equal to $N_1 N_2 N_3$, the number of unit cells in the extended cell. Since the total number of nuclei is an integral multiple of this, so is the number of electrons. It follows, if the bands do not overlap and the number of electrons is even, that all the occupied bands will be exactly filled with electrons. On the other hand, if the number of electrons is odd the lower occupied bands will be completely filled but the upper occupied band will be half filled, up to what is called the Fermi level. With filled bands the system will be an insulator (or an intrinsic semiconductor) for a finite energy, equal to a band gap, will need to be supplied before there is any possibility of charge transport. (In an intrinsic semiconductor the gap energy is bridged by thermal excitations, so the conductivity falls as the temperature drops.) With a half-filled band there are plenty of empty nearby energy states, so an almost negligible amount of energy is needed to excite an electron from just below the Fermi level to just above it. Such systems show electrical conduction. In other cases the bands overlap and the higher-lying occupied bands are not filled even with an even number of electrons.

In the presence of an external magnetic field it is often assumed that all orbital moment is quenched. Since most occupied band states contain two electrons with opposite spins, it is then convenient to regard each band as doubled, one with + spins relative to the field direction and the other with − spins. The effect of the field is to raise one band relative to the other, so a lower-energy state can be reached by having electrons near the Fermi level of the raised band transfer to empty states just above the Fermi level of the other band. The new ground state has a net magnetic moment in the direction of the applied field and so has a paramagnetic moment. To determine its variation with temperature it is necessary to take account of thermal excitations of electrons into empty states above the new ground state. The calculation is fairly complicated and is best carried out using the grand partition function and a special technique, known as second quantization, neither of which have yet been described. It shows that the paramagnetism is field and temperature independent, as is found in most metals. [For an introduction to quantum statistical mechanics see Toda et al. (1983). The assumption that there is no induced orbital moment is not really valid, nor is the statement that metals have temperature-independent susceptibilities, for much depends on the range of temperature in use and the purity of the sample. Studies of the magnetic properties of very pure metals at low temperatures have led to the discovery of a range of unusual properties and the need for even more elaborate theory. An interested reader can gain entry to the literature through references to Landau levels.]

For the ferromagnetic metals the approximation of replacing the Coulomb interactions between the electrons by an interaction in which each electron is in a periodic potential and then filling the bands up to a Fermi level is clearly not good enough, for there is no obvious way of producing macroscopic moments of the observed magnitudes. Nevertheless, a treatment through band theory is attractive because it does spread the spin of each electron over the whole lattice and so gives a way of explaining why there will be nonintegral spin moments at lattice sites when the plus and minus spin bands are relatively displaced. The problem is to demonstrate that including the Coulomb interactions enhances this effect in the ferromagnetic metals.

§ 5.8 Crystal Field and Band Theory Compared

There is really very little difference between the basic Hamiltonians of magnetic insulators and those of conductors, just a different set of nuclear charges and a different number of electrons. Also, the initial approximation, the replacement of the Coulomb interactions between the electrons by an effective potential, or potentials, to give one-electron Hamiltonians, is common to both. However, when the theory for a magnetic insulator is examined in more detail, it is clear that this is not what has actually been done, for it is only the Coulomb interactions of an ion with it immediate neighbors that have been replaced by an effective potential. In consequence, part of the Coulomb interaction has not been replaced by an effective potential and the part which has been has resulted in the elimination of the periodicity of the lattice and the violation of the indistinguishability of electrons. When expressed this way the whole procedure looks questionable, but against that it should be remembered that it is an approximation and perturbation theories are available to improve on approximations. Nevertheless, having violated a rigorous symmetry requirement it is difficult to restore it subsequently. What has then happened is that crystal field theory has led to spin-Hamiltonians for individual ions which have then been supplemented by further spin-Hamiltonian-like interactions between the ions that seem to restore the lattice periodicity. It is much less obvious that the requirement that electrons should be indistinguishable has simultaneously been met. On both scores, indistinguishability and periodicity, band theory appears to have the edge, except that it is difficult to go beyond the effective-potential approximation. As a result, it has not got very far with the description of magnetic insulators.

The next few chapters will be devoted to showing that band theory is not essential to the treatment of periodic systems, that indistinguishability between electrons can be retained without it, that it is then possible to arrive at the spin-Hamiltonians deduced for isolated magnetic ions through crystal field theory, and that the interactions between them which have largely been introduced phenomenologically emerge naturally. The path is quite long and in one sense unrewarding, for it arrives at effective spin-Hamiltonians that have a familiar form and which have been amply confirmed as correct by experiment. Nevertheless, there is some satisfaction in having a more viable derivation. Fortunately, there is more, for the new way looks much more directly at many-electron systems and it is then found that it is possible to use the same approach quite successfully for some other magnetic systems that present insuperable difficulties for crystal field theory and major difficulties for band theory. There is also some reason to hope that the new method will help to resolve the question of whether to use localized or delocalized models. An account of these extensions will necessarily come after the demonstration of the improved derivation of the existing spin-Hamiltonians.

§ 5.9 References

Anderson, P.W. 1963. In *Magnetism*, ed. G.T. Rado and H. Suhl (Academic Press: New York), vol. 1, p. 25.

Bates, L.F. 1961. *Modern Magnetism* (University Press: Cambridge).

Bitter, F. 1937. *Phys. Rev.* 54:79.

Bizette, H. et al., 1938. *Compt. Rend.* 207:449.

Bleaney, B., and Bowers, K.D. 1952a. *Phil. Mag.* 43:372.

Bleaney, B., and Bowers, K.D. 1952b. *Proc. Roy. Soc.* A214:451.

Caspers, W.J. 1989. *Spin Systems* (World Scientific: Singapore).

Curie, P. 1895. *Ann. Chim. Phys.* 5:289.

Dirac, P.A.M. 1926. *Proc. Roy. Soc.* A112:661.

Furrer, A. 1977. *Crystal Field Effects in Metals and Alloys* (Plenum: New York).

Gallison, P. 1987. *How Experiments End* (University of Chicago Press: Chicago), Chap. 2.

Griffiths, J.H.E. 1951. *Physica* 17:253.

Heisenberg, W. 1926. *Zeits. Phys.* 38:411.

Hudson, R.P. 1972. *Principles and Applications of Magnetic Cooling* (North-Holland: Amsterdam).

Judd, B.R. 1963. *Operator Techniques in Atomic Spectroscopy* (McGraw-Hill: New York).

Langevin, P. 1905. *Ann. Chim. Phys.* (Paris) 5:70.

Mattis, D.C. 1985. *Theory of Magnetism 2* (Springer-Verlag: Berlin).

Néel, L. 1932. *Ann. Phys.* (Paris) 18:5.

Oitmaa, J., Hamer, C.J., and Zheng, Weihong. 1994. *Phys. Rev.* B 50:3877.

Owen, J., and Stevens, K.W.H. 1953. *Nature, Lond.* 171:836.

Shull, C.G. et al. 1951. *Phys Rev.* 83:333.

Slater, J.C. 1968. *Quantum Theory of Matter* (McGraw-Hill: New York).

Stevens, K.W.H. 1953. *Rev. Mod. Phys.* 25:166.

Toda, M., Kubo, R., and Saito, N. 1983. *Statistical Physics I* (Springer-Verlag: Berlin).

Van Vleck, J.H. 1935. *J. Chem. Phys.* 3:807.

Van Vleck, J.H. 1937a. *J. Chem. Phys.* 5:320.

Van Vleck, J.H. 1937b. *Phys, Rev.* 52:1178.

Weiss, P.R. 1907. *J. de Phys.(4)* 5:70.

Weiss, P., and Foëx, G. 1931. *Le Magnétisme* (Libraire Armand Colin: Paris).

§ 5.10 Further Reading

There are a number of publications that list experimentally determined quantities, with the Landolt-Börnstein series (*Zahlenwerte und Funktionen*, Springer-Verlag: Heidelberg) being particularly useful. Data about ionic energy levels can be found in I Band, I Teil (1950). There is also a series of publications under the title *Atomic Energy Levels*, published by the National Bureau of Standards from 1949 onwards. There are also two publications, both entitled *Handbook of Atomic Data*, one dated 1976, by J. Karwowski and another, dated 1979, by S. Fraga, J. Karwowski, and K.M.S. Saxena. Both are published by Elsevier: Amsterdam.

Group Theory in Solid State Physics by A.P. Cracknell (1975) (Taylor and Francis: London) gives a detailed account of the application of group theory to the study of periodic lattices.

6

Second Quantization

§ 6.1 The Concept of Second Quantization

It will be assumed that no eigenfunctions of any many-electron Hamiltonian are known exactly, which indicates that the task of finding even one is difficult. When this is compounded with the physical requirement that the only ones of physical interest have to be antisymmetric with respect to interchanges of electrons the task looks even more daunting, for how are these to be picked out ab initio? There is therefore interest in any technique that concentrates attention on functions that already incorporate the antisymmetric requirements, which is why Slater determinants were introduced. It must be admitted, though, that they are clumsy to write down and manipulate, so a more recent development, which modifies the Hamiltonians so that in their new forms all the eigenfunctions are automatically antisymmetric, has much to recommend it. The modification, which can also be developed if the eigenfunctions are required to be symmetric under interchanges of particles, a condition which is imposed on particles with integer spins, is known as "second quantization."

The method begins by assuming that a complete set of orthonormal one-electron wave functions is available, which is a problem in itself, for such sets are quite rare. For example, the eigenstates of the hydrogen atom with the nucleus at a fixed position might seem to provide such a set. There is, though, a problem over the high-energy eigenstates, which describe unbound states. It then might seem best to assume that at a sufficiently large radial distance there is a barrier of infinite height, which will not significantly alter any of the low-lying eigenstates and will ensure that all the eigenstates can be normalized. The eigenstates of a three-dimensional harmonic oscillator, all of which are bound, form a set where there is no such problem. In each of these examples, if the functions are to be used to describe electrons, each orbital function must be multiplied by one or other of two spin kets, $|+\rangle$ and $|-\rangle$.

Suppose now that an orthonormal set $[\phi_n]$ is available, where spin is included in the definition of each ϕ_n. For a two-electron system a typical Slater determinant would be of the form $\{\phi_p, \phi_q\}$, where the p and q subscripts refer to states in the set, not to the electrons. It then becomes possible to define an operator a_r^* which, when applied to this determinant, changes it into another Slater determinant, that of a three-electron system, with ϕ_r appearing in the right-hand position. That is, $a_r^*\{\phi_p, \phi_q\}$ gives $\{\phi_p, \phi_q, \phi_r\}$. (It should be re-called that the definition of a Slater determinant includes a normalizing factor. The definition of a_r^* includes the requirement that the added state is to the right of the states already present and that the normalizing factor is appropriately adjusted.) Similar operators, a_n^*, can be defined, one for each state in the basis set. They can each be applied to any of the 2×2 Slater determinants $\{\phi_p, \phi_q\}$, though a 3×3 Slater determinant will only be obtained if the operator gener-ates a state that is not already present in the determinant. If an operator appears to generate a state that is already present (occupied) the new determinant has two identical columns and so vanishes. The next step is to define unstarred operators, denoted by a_n, one for each ϕ_n. Their role is to remove states from Slater determinants. It is convenient to begin with a 3×3 example. A fur-ther convention is obviously needed, to specify where the state which is to be removed is situated in the $\{\dots\}$ determinantal form. The definition therefore states that before a_n can operate the state ϕ_n has to be moved to the far right by interchanging columns, if necessary, a procedure which may result in the sign of the determinant being reversed. The application of a_n then removes the ϕ_n and adjusts the normalizing factor. If ϕ_n is not present the action of a_n is defined as giving zero. The final step is to extend the definitions of the starred (creation) and unstarred (annihilation) operators so that they apply to Slater determinants of any size, including those that have only one row and column (which would not normally be described as determinants).

The operators just defined satisfy very simple relations which are usually known as anticommutation rules, for

$$a_m a_n^* + a_n^* a_m = 0$$

if ϕ_n and ϕ_m are different one-electron states, and

$$a_n a_n^* + a_n^* a_n = 1$$

if they are the same. For two starred or two unstarred operators the anticom-mutation rule has the $+$ sign on the left of the equation and *always* a zero on the right-hand side. That is,

$$a_m^* a_n^* + a_n^* a_m^* = 0 \text{ and } a_m a_n + a_n a_m = 0.$$

(Similar rules, except that the plus signs between the pairs on the left are re-placed by minus signs, hold for the creation and annihilation operators used in describing harmonic oscillators and for particles which obey Bose statistics.) The convention is that the rules are described as commutation when the minus signs are there and anticommutation when the plus signs are there. However, unless both Fermi and Bose particles are present the use of "anti" is unnecessary. So in the following pages "commutation" will be used in place of "anticom-mutation." (This does not mean that the rules have been altered, but just that a change has been made in the phrase used to refer to them.)

It now becomes possible to express a sum of one-electron operators, taken between two determinantal wave functions, as an expression in matrix elements and annihilation and creation operators. As already shown (§1.9) any one-electron operator has a vanishing matrix element between any two determinants that differ in more than one row and column. Consider therefore the expression

$$\sum_{n,m} \langle n|V_1|m\rangle a_n^* a_m, \tag{6.1}$$

where V_1 represents a typical one-electron operator, such as would appear summed over all electrons in a normal Hamiltonian. (The subscript 1 does not denote electron 1 but simply that the operator is a one-electron operator, such as a typical kinetic energy expression, $\mathbf{p}^2/2m$.) Now compare the matrix element of the summation over all electrons of V_1 with that of (6.1) when they are taken between the same two determinantal states. If the determinants differ in two one-electron states the former gives zero, as has been shown in §1.9. Each $a_n^* a_m$ in (6.1) also gives zero, for all such an operator does is to replace ϕ_m, if it is present in the determinant on the right, by ϕ_n, and so leave a determinantal state on the right which is orthogonal to the one on the left. If the determinants differ by only one state the summed V_i gives $\langle p|V_1|q\rangle$, where $|p\rangle$ and $|q\rangle$ are the two different states. Looking at the $a_n^* a_m$ operator pairs, it is seen that they all have zero matrix elements except when $n = p$ and $m = q$, for it is only for this choice that the state ϕ_q in the determinant on the right is changed into ϕ_p. (6.1) has been defined so that this operator pair occurs with the right coefficient, $\langle p|V_1|q\rangle$, to ensure that (6.1) reproduces the matrix elements of V_1 summed over electrons between all determinantal states that differ either by one or two one-electron states. The step that shows that they also agree when the two determinantal states are identical follows with a similar argument. Thus (6.1) is equivalent to V_1 summed over electrons. That is,

$$\sum_{i=1}^{N} V_1(i) \equiv \sum_{n,m} \langle \phi_n|V_1|\phi_m\rangle a_n^* a_m. \tag{6.2}$$

On the left the operator V_i has a subscript 1 to indicate that it is a one-electron operator and the summation is over N electrons. On the right the matrix element is one which would occur for the same V in a problem involving just one electron. (There is really no need to give it the subscript for the $\langle n|V|m\rangle$ form indicates that it is a one-electron operator.) The summation is over all operator pairs, $a_n^* a_m$, where each operator is defined in terms of the one-electron states of a complete orthonormal basis set, $\{\phi_n\}$.

On taking the matrix elements between any two $N \times N$ determinantal states, the two sides of (6.2) give identical results. That does not imply, however, that the two sides are identical, which is why they have been related by an equivalence, $\equiv$, rather than an equality symbol. The expression on the right does not depend on N, the number of electrons, whereas the one on the left is usually regarded as describing a system with a definite number of electrons. So the right hand one actually operates in a much larger Hilbert space, the space of all numbers of electrons. But since each starred operator is paired with an unstarred operator, the pair has no matrix elements between states with differing numbers of electrons.

By similar reasoning it is possible to recast the two-electron operators, summed over all electrons, such as are present in the generic and many model Hamiltonians, in a second quantized form. Thus:

$$\tfrac{1}{2} \sum_{i \neq j} V_2(i, j), \tag{6.3}$$

where the subscript 2 on V indicates that it is a two-electron operator and i and j label different electrons, can be replaced by

$$\tfrac{1}{2} \sum_{n,m,p,q} \langle n, m | V_2 | p, q \rangle a_n^* a_m^* a_q a_p, \tag{6.4}$$

where V_2 is the two-electron operator that would appear as a coupling between the electrons in a related system in which the number of electrons had been reduced to 2. It should be noted that the order n, m, p, q in the bras and kets of the matrix element is different from that in the sequence of operators, being replaced by n, m, q, p. (This is necessary because in the bras and kets n and p refer to the first electron and m and q to a second. In changing from a bra to a related ket the first symbol in both refers to the same electron.) Also, because of the summation over all p, q, r, and s, a given combination of operators, such as $a_n^* a_m^* a_q a_p$, may occur several times in a disguised form. For example, with $p \neq q$ the commutation rules can be used to show that $a_n^* a_m^* a_p a_q$ is the same as $-a_n^* a_m^* a_q a_p$. While the latter is simply the negative of the former, the matrix element which multiplies $a_n^* a_m^* a_q a_p$ is unlikely to be equal to that which multiplies $a_n^* a_m^* a_p a_q$.

Operators in which three or more electrons are coupled are not present in the generic Hamiltonian, nor are they present in most model Hamiltonians. So the standard assumption is that Hamiltonians contain only one- and two- electron operators. The form

$$\sum_i H_1(i) + \tfrac{1}{2} \sum_{i \neq j} H_2(i, j), \tag{6.5}$$

where H_1 includes all the one-electron and H_2 all the two-electron operators in the Hamiltonian, can be replaced by

$$\sum_{n,m} \langle n | H_1 | m \rangle a_n^* a_m + \tfrac{1}{2} \sum_{n,m,p,q} \langle n, m | H_2 | p, q \rangle a_n^* a_m^* a_q a_p. \tag{6.6}$$

(The suffixes on H_1 and H_2 are not really needed, because the form of each bra and ket identifies the operator as being of one- or two-electron type.) It is known as a second quantized form of the original Hamiltonian, which from now on will be referred to as the first quantized Hamiltonian. It may be noted that in (6.5) i and j label electrons and that in (6.6) the subscripts 1 and 2 identify the operators as being either of one- or two-electron type and, as with the one-electron example, the relationship between (6.5) and (6.6) is an equivalence.

In setting up the second quantized form a complete set of one-electron orthogonal functions has been invoked without an explicit statement of what functions it actually contains. It follows that the *relationship* between (6.5) and (6.6) does *not depend* on which complete basis set has been chosen. This implies that most of the properties of the second quantized form, (6.6), do not depend on the choice of the orthonormal basis set, $\{\phi_n\}$. For any choice it will have the same eigenvalues and associated degeneracies. ("Most" rather than "all" has been used because with a degenerate set of eigenstates those found with one basis set are unlikely to come out to be identical with those found with another.)

The property of a second quantized Hamiltonian, that it appears to describe all possible numbers of electrons at once, takes a little getting used to. One thing it leads to is a need to modify the way in which such Hamiltonians are treated in statistical mechanics (see §6.2). Then there is a possible source of confusion over the use of "matrix element." This is usually taken to describe an expression such as $\langle n|V|m \rangle$ in which an operator V is sandwiched between a bra and a ket vector. But when (6.6) is examined it is obvious that it already contains matrix elements, those of a different but related system, and that additional ones will be produced if the operator is sandwiched between a bra and a ket vector. Usually the meaning of "matrix element" is clear from the context in which it is being used, so there is no source of confusion. But where there may be, "matrix element" will be reserved for the related Hamiltonian, so it is a coefficient that multiplies a one- or two-electron operator in a second quantized Hamiltonian. In the few cases where the ket and bra describe a system with a much larger number of electrons the different meaning of matrix element will be pointed out.

The second quantized Hamiltonian of a particular atom describes it however many electrons it has, so it describes all its states of ionization. Any one of these can be described by restricting the Hilbert space to that of a fixed number of electrons. This can be done by introducing the concept of a vacuum state, a single state, usually denoted by $|v\rangle$, which contains no electrons and which is such that any annihilation operator applied to it gives zero. If, however, a creation operator is applied the state which results (or is created) is identified as having one electron in the state that has been used in the definition of the creation operator. (There is no concept of antisymmetry when only one electron is present.) If two different creation operators are applied, to give $a_n^* a_m^* |v\rangle$, the Slater determinant $\{m, n\}$ is obtained, where the ordering of n and m is important. The first operator has produced $\{m\}$ (created an electron in ϕ_m). The second has created another, in ϕ_n, to give a Slater determinant, $\{\phi_m, \phi_n\}$, in which n is in a column to the right of m. If the operators had been applied in a reversed order the same physical state would have been obtained, though the Slater determinant would have the opposite sign. If the annihilation operator a_n is applied to $\{\phi_n, \phi_m\}$ it is first necessary to interchange ϕ_n and ϕ_m so that ϕ_n is to the right of ϕ_m and then remove it by the action of a_n, to leave $-\{\phi_m\}$.

These kinds of manipulation become increasingly tedious as the number of electrons is increased, so it is an attractive feature of the second quantized formalism that they are seldom necessary. An alternative to such a sequence of operations is provided by the commutation rules. For example, the same sequence can be written

$$a_n a_m^* a_n^* |v\rangle = -a_m^* a_n a_n^* |v\rangle = -a_m^* (1 - a_n^* a_n)|v\rangle = -a_m^* |v\rangle,$$

where the basic technique is to use the commutation rules to move annihilation operators to the right. Two possibilities present themselves. Either the corresponding creation operator is encountered or it is not. If it is not then the annihilation operator can be moved to the far right when it acts on the vacuum state to give zero. There is therefore no need to take account of any minus signs. If, however, it does arrive to the immediate left of its creation operator then the aa^* pair can be replaced by $1 - a^*a$ and the a^*a part can be dropped because the unstarred operator in it can be moved to be adjacent to the vacuum state, to give zero. The product $a_n a_n^*$ is therefore equivalent to unity.

It may be noted that the form of the second quantized Hamiltonian given in (6.6) has all the annihilation operators to the right of the creation operators. This is the so-called canonical form. It is never applied to the vacuum state, for it would simply give zero. Instead, for an N-electron system, it is applied to states that have been obtained by applying products of N different creation operators to the vacuum state.

In some cases it is assumed that to a good approximation the Hilbert space can be restricted. For example, with a many-electron atom the inner shells are usually regarded as filled in the unperturbed ground state and a common assumption is that they remain filled under some perturbation. The vacuum state can then be taken as a many-electron state in which the closed shells are already filled. The second quantized Hamiltonian can then be truncated so that it contains no annihilation operators that would remove electrons from the filled shells. Similarly, with transition metal ions it is sometimes convenient to restrict the Hilbert space to one in which the number of electrons in the open (partially filled) shell is such that it is more than half filled. The vacuum state can be chosen to be the filled shell, with annihilation operators applied to it to reduce the number of electrons to that required. Since the commutation rules are unaltered if every a^* operator is replaced by an unstarred operator, b, and vice versa, it then becomes possible to introduce "hole" creation and annihilation operators. A state with $N-1$ electrons in a shell that is filled with N electrons can be regarded as created by a hole operator acting on a vacuum state that contains no holes.

§ 6.2 Statistical Modifications

There is often an interest in the thermodynamic properties of a solid and in many cases it is convenient to assume that in thermal equilibrium there are small interchanges of energy with a thermal bath, but no interchanges of electrons or nuclei. Statistical mechanics has then shown that the partition function, Z, defined as the trace of $\exp(-H/kT)$, is an important concept, because all the thermodynamic properties can be obtained from it by appropriate manipulations. In elementary examples it is usually further assumed that all the eigenstates of H are of physical significance. Such an assumption is not necessarily valid for many-electron systems because a given Hamiltonian may have eigenstates that violate the exclusion principle. So their eigenvalues should be omitted in evaluating Z. As H in a second quantized form has already eliminated any nonphysical states it would seem sensible to use this form of H in evaluating Z. There is, however, a complication, that the number of electrons appears to have become indefinite, a feature that is not necessarily undesirable for there may indeed be interchanges of electrons between the system of interest and a thermal bath that is determining its temperature. So the question of the appropriate definition of Z arises. The answer is that the second quantized Hamiltonian should be modified by the addition of an extra term, $-\mu\mathcal{N}$, where μ is a scalar variable known as the chemical potential and $\mathcal{N}$ is an operator defined as

$$\mathcal{N} = \sum_n a_n^* a_n. \tag{6.7}$$

It is called the number operator because if it is applied to any state obtained by applying a sequence of creation operators to the vacuum state it multiplies the state by the number of creation operators, the number of electrons. So, effectively, the operator sums the electrons. The modified Hamiltonian becomes

$$\sum_{n,m} \langle n|H_1|m\rangle a_n^* a_m - \mu \sum_n a_n^* a_n + \frac{1}{2} \sum_{n,m,p,q} \langle n,m|H_2|p,q\rangle a_n^* a_m^* a_q a_p. \tag{6.8}$$

As far as eigenvalues and eigenstates are concerned the additional term is of no physical significance, for with a fixed number of electrons all the addition has done has been to displace all energy levels by the same amount. The changes come in statistical mechanics. The partition function, Z, has to be replaced by what is know as the grand partition function (GPF). It equals the trace of $\exp(-H'/kT)$, where H' is the modified Hamiltonian given in (6.8). Writing β in place of $1/kT$ the Helmholtz free energy, F, is replaced by the Gibbs free energy, defined as $G = -(1/\beta)\log$ (GPF). The value of μ and the other thermodynamic properties are then obtained from

$$N = -\frac{\partial G}{\partial \mu}, \quad S = -\frac{\partial G}{\partial T}, \tag{6.9}$$

where N is the mean number of electrons. [The number operator, $\mathcal{N}$, is sometimes replaced by N, which is questionable. In (6.8) it is an operator and it is incorrect to replace it by a scalar except in the context of statistical mechanics when a definite determination of the value of μ has occurred.]

A commonly used example in band theory is one in which the model Hamiltonian is taken to have the form

$$H = \sum_{\mathbf{k}} (\epsilon_{\mathbf{k}} - \mu)a_{\mathbf{k}}^* a_{\mathbf{k}}, \tag{6.10}$$

where $\mathbf{k}$ is the propagation vector for a Bloch wave function, the totality of which has provided the set which defines the creation and annihilation operators. Since $a_{\mathbf{k}_1}^* a_{\mathbf{k}_1}$ commutes with $a_{\mathbf{k}_2}^* a_{\mathbf{k}_2}$ when $\mathbf{k}_1 \neq \mathbf{k}_2$ the Hamiltonian is a sum of commuting terms, each of which can be regarded as operating in its own Hilbert space. The trace of $\exp(-\beta H)$, which should be evaluated using the states in the Hilbert space of the whole Hamiltonian, can therefore be written as a product of the traces of the commuting parts, taken over their respective Hilbert spaces. Thus the term in $a_{\mathbf{k}}^* a_{\mathbf{k}}$ contributes a factor

$$\text{Trace } \{\exp[-\beta(\epsilon_{\mathbf{k}} - \mu)a_{\mathbf{k}}^* a_{\mathbf{k}}]\}$$

and $a_{\mathbf{k}}^* a_{\mathbf{k}}$, being a number operator for a single electron, takes just two values, 0 and 1. The GPF is therefore the product of factors of the form

$$[1 + \exp\{-\beta(\epsilon_{\mathbf{k}} - \mu)\}].$$

(6.9) then leads to an implicit equation

$$N = \sum_{\mathbf{k}} \frac{1}{1 + e^{\beta(\epsilon_{\mathbf{k}} - \mu)}} \tag{6.11}$$

which relates μ to the mean number, N, of electrons. The expression

$$\frac{1}{1 + e^{\beta(\epsilon_{\mathbf{k}} - \mu)}} \tag{6.12}$$

is known as the Fermi distribution function. It can be interpreted as giving the mean number of electrons in the one-electron state $|\mathbf{k}\rangle$, which has energy $\epsilon_{\mathbf{k}}$. If $T \to 0$, $\beta \to \infty$, so the distribution function becomes 1 if $\epsilon_{\mathbf{k}} < \mu$ and 0 if $\epsilon_{\mathbf{k}} > \mu$. The value of μ, when $T = 0$, is called the Fermi level, or the chemical potential at absolute zero. [In using this distribution function it should be recalled that in the definition of the basis states the spin orientations are included. In many accounts $|\mathbf{k}, +\rangle$ and $|\mathbf{k}, -\rangle$ are used to separate the spin possibilities. If so, $|\mathbf{k}\rangle$ then describes just the orbital part of the wave function and a factor of 2 should be included in the numerator of (6.12), on the assumption that the two spin states of the same $\mathbf{k}$ are degenerate in energy.]

§ 6.3 Choice of Basis States

The use of second quantized Hamiltonians is so common that many authors seem to be under the impression that it is unnecessary to give the basis states that define the creation and annihilation operators, which is surprising because suitable basis sets are not all that easily obtained. The eigenstates of a free particle Hamiltonian, $\mathbf{p}^2/2m$, are not convenient because of the normalization problems, and even if they were they would seem to be an unpromising choice for solid state problems. A better choice might seem to be to use the functions that describe the motion of an electron when it is moving in a periodic potential, except that exact solutions of such problems are seldom available. Nevertheless, some progress is possible, for with periodic boundary conditions the normalization can be taken over a finite volume, the extended cell, the eigenvalues form bands the states of which can be characterized by $\mathbf{k}$ vectors, and there is the certainty that states with differing $\mathbf{k}$ values are orthogonal. A basis state for the second quantization can therefore be labeled with an index n to denote the band it is in, a $\mathbf{k}$ vector to identify it within the band, and an index which is either $+$ or $-$ to denote the spin orientation, the eigenvalues of s_z, using some convenient direction for Oz.

In many papers the one-electron part of the second quantized Hamiltonian is assumed to take the form

$$\sum_{n,\mathbf{k},\sigma} \epsilon_{n,\mathbf{k}} a^*_{n,\mathbf{k},\sigma} a_{n,\mathbf{k},\sigma} \qquad (6.13)$$

with the implication that the $\epsilon_{n,\mathbf{k}}$ are the energies of the band states. There is then a problem, because in most discussions of band theory it is assumed that the one-electron periodic potential is an effective potential that includes not only the potential due to the lattice of nuclear charges but also a time-independent average potential due to all the other electrons. In a metal the two should almost cancel. But from the point of view of second quantization the Coulomb interactions between the electrons are genuine two-particle interactions and so should appear in the two-particle a^*a^*aa part of the second quantized form. A band structure based on an effective periodic potential will produce Bloch functions of just the form needed in setting up the second-quantization formalism. On the other hand, it seems unavoidable that if these are used as a basis the $\epsilon_{n,\mathbf{k}}$ values in its one-electron part will be the expectation values of the one-electron part of the generic Hamiltonian taken over states that have been determined using an approximation which has taken some account of the two-electron interactions. This makes it difficult to known how to deal with the second quantized part of the two-electron interactions. The origin of the difficulty is that second quantization treats the one- and two-electron parts of the initial Hamiltonian as separate entities, whereas in the usual band theories they are muddled together. Looked at another way, the one-electron part gives a sum of operators each of which is a product of one starred and one unstarred operator, while the two-electron part gives a sum of operators each of which is a product of two starred and two unstarred operators. There is no obvious way in which the latter can be replaced, either partly or entirely, by the former.

To be specific, and in an attempt to avoid confusion, it will be assumed that the Bloch functions are those that have emerged from a theory which has used a one-electron periodic potential such as is likely to be used in a typical band structure calculation; one which does its best to simulate the two-electron Coulomb interactions by a one-electron potential. The $\epsilon_{n,\mathbf{k}}$, the one-electron coefficients, will therefore not be regarded as band energies but simply as one-electron matrix elements. Similarly, the two-electron coefficients will be well defined. There is then no inconsistency if a change of basis is made, because of the invariance property of a second quantized form. It is only when approximations are made that errors may arise.

§ 6.4 Wannier Functions

Another set of basis functions is provided by the Wannier functions (Wannier, 1937). They can be regarded as the Fourier transforms of the Bloch functions. Their definition uses the concept of bands and is quite general. It will, though, be introduced in the context of a simple cubic lattice of identical nuclei. A lattice that extends to infinity causes some inconvenience so it will be regarded as divided into extended cells each of which contains N^3 unit cells, where N is a large number. The purpose of the extended cells is to allow the introduction of additional symmetry, the assumption that the *wave functions* in the one-electron potential, which has the symmetry of the simple cubic lattice, can be taken as periodic in the much larger lattice formed by the extended cells, which will be referred to as the extended lattice. It follows that three translation operators, T_x, T_y, and T_z, which respectively represent displacements by one unit cell in the x, y, and z directions, can be introduced. After N repeated displacements any wave function will have returned to its original value. So each operator satisfies $T^N = E$ and together they can be used to generate a finite group of commuting operators. The irreducible representations of the group can be denoted by **k** vectors, with a typical i.r. having three components, (k_x, k_y, k_z), where $k_x = 2\pi p/N$, $k_y = 2\pi q/N$, and $k_z = 2\pi r/N$, where p, q, and r are integers (see §5.7). It follows that every normalized eigenfunction of the Hamiltonian which describes the motion of a particle in the periodic potential can be chosen to be a generator of a one-dimensional i.r. It will have a typical form:

$$\psi_{n,\mathbf{k}}(r) = \exp(i\mathbf{k} \cdot \mathbf{r})u_{n,\mathbf{k}}(\mathbf{r}), \tag{6.14}$$

where **r** denotes a point in the extended cell and $u_{n\mathbf{k}}(\mathbf{r})$ is a function that has the periodicity of the lattice, n being a band index. [This result is known as Bloch's theorem. It can be derived in several ways; see Ziman (1964), §1.4, for example. Once the factor $\exp(i\mathbf{k} \cdot \mathbf{r})$ has been extracted, $u_{n\mathbf{k}}(\mathbf{r})$ is necessarily periodic if the wave function is to be a basis element of an i.r.] The set of functions generated by varying n and **k** is similar to that introduced in the tight-binding model, though there is an important difference. Beginning with a Hamiltonian ensures that all its eigenfunctions are mutually orthogonal, so no further orthogonalizations of functions with the same **k** are needed. Each has been assumed to be normalized to unity by the integration of its square modulus over the volume of the extended cell.

The Wannier functions are then defined by

$$W_{n,\mathbf{R}_i}(\mathbf{r}) = \frac{1}{N^{3/2}} \sum_{\mathbf{k}} \exp(-i\mathbf{k} \cdot \mathbf{R}_i)\psi_{n,\mathbf{k}}(\mathbf{r}), \tag{6.15}$$

where $\mathbf{R}_i$ is a vector from an origin centered on an ion at a vertex of the unit cell at the center of the extended cell to an equivalent ion in the ith unit cell. The function is normalized to unity by integration over the extended cell, which contains N^3 unit cells. The orthogonality of the $\psi_{n\mathbf{k}}$'s ensures that the Wannier functions with different n's are orthogonal. So also are the Wannier functions with the same n and different $\mathbf{R}_i$'s, for in the integration over **r** of

$$\sum_{\mathbf{k}_1,\mathbf{k}_2} \exp[-i(\mathbf{k}_1 \cdot \mathbf{R}_i - \mathbf{k}_2 \cdot \mathbf{R}_j)]\psi^*_{n,\mathbf{k}_1}(r)\psi_{n,\mathbf{k}_2}(\mathbf{r}) \tag{6.16}$$

all the terms with $\mathbf{k}_1 \neq \mathbf{k}_2$ vanish, to leave a summation

$$\sum_{\mathbf{k}_1} \exp[-i\mathbf{k}_1 \cdot (\mathbf{R}_i - \mathbf{R}_j)] \tag{6.17}$$

which vanishes unless $\mathbf{R}_i = \mathbf{R}_j$. So the Wannier functions form an orthonormal set. In addition, the set defined using one choice of origin is identical with that found with any equivalent origin. This can be shown by comparing the Wannier function with a given n when $\mathbf{R}_i$ is taken to be zero with the Wannier function which has the same n but which is defined for the cell at $\mathbf{R}_i$. The former is

$$W_{n,0}(\mathbf{r}) = \frac{1}{N^{3/2}} \sum_{\mathbf{k}} \psi_{n,\mathbf{k}}(\mathbf{r}) \tag{6.18}$$

and the latter is

$$W_{n,\mathbf{R}_i}(\mathbf{r}) = \frac{1}{N^{3/2}} \sum_{\mathbf{k}} \exp(-i\mathbf{k} \cdot \mathbf{R}_i)\psi_{n,\mathbf{k}}(\mathbf{r}). \tag{6.19}$$

If $\mathbf{r}$ in (6.19) is replaced by $\mathbf{R}_i + \mathbf{s}$, which amounts to moving the origin of coordinates to the cell at $\mathbf{R}_i$, the relation

$$W_{n,\mathbf{R}_i} = \frac{1}{N^{3/2}} \sum_{\mathbf{k}} \exp(-i\mathbf{k} \cdot \mathbf{R}_i)\psi_{n,\mathbf{k}}(\mathbf{R}_i + \mathbf{s}) = \frac{1}{N^{3/2}} \sum_{\mathbf{k}} \psi_{n,\mathbf{k}}(\mathbf{s}) \tag{6.20}$$

shows that the Wannier function for $\mathbf{R}_i$ is exactly the same as that for the origin. Thus the Wannier functions form a set of orthonormal functions which can be regarded as divisible into N^3 families, one for each cell of an extended cell, with the family in one lattice cell being identical, apart from its position, with the family in any other cell. (Each Wannier function repeats itself with the periodicity of the extended cell structure, so if N is sufficiently large it can be expected that a Wannier function that is large in the vicinity of a particular unit cell will decrease in magnitude outside the cell and eventually become so small that the error in assuming that it is normalized to unity is negligible. This may be incorrect, particularly for the Wannier functions of high-lying bands.)

The eigenstates of a hydrogenic problem can be found exactly: they form a denumerable infinity of bound states, with an innumerable number of unbound states. In tight-binding theory each band is based on a localized function, which leads, before orthogonalizing, to states of the same $\mathbf{k}$ in different bands with their description as $(1s)$, $(2s)$, $(2p_x)$, etc., bands. There is the implication that after orthogonalizing, a step which then allows the construction of Wannier functions, the same description can be used. The number of Wannier functions per band is the same as the number of cells per extended cell. It is not obvious how each of these functions can be put into 1:1 correspondence with an eigenstate of hydrogen. The best that can be expected is that with a properly chosen periodic potential the low-lying Wannier functions, those that are similar to the $(1s)$, $(2s)$, etc. hydrogenic functions, form a good enough complete set.

They will have been constructed from Bloch functions which have differing $\mathbf{k}$ vectors, any one of which can be given an arbitrary phase factor, showing that the Wannier functions formed from a particular band are by no means unique.

Indeed, they do not even need to be constructed from Bloch functions in the same band. (That there will be a good deal of ambiguity is to be expected, for the members of any orthonormal set can be combined in many different ways to give other orthonormal sets.) The ambiguity opens the way to creating Wannier functions with special symmetry properties.

The Wannier functions formed from a particular band all contain the Bloch function that has $\mathbf{k} = \mathbf{0}$, a function which has an identical shape in the vicinity of every lattice point. Its transformation under the point group will therefore give a representation of the group and there is no loss in generality in assuming that the representation is an i.r. Suppose, therefore, that it happens to have p_x-like symmetry. The question then arises as to whether it is possible to arrange the summation over all the other $\mathbf{k}$'s so that it too has p_x-like symmetry. This is where the flexibility in the definition of Wannier functions is useful. A Bloch function with an arbitrary $\mathbf{k}$ can be related to a set of $\mathbf{k}$'s, known as the star of $\mathbf{k}$, by the operations of the point group as they change one $\mathbf{k}$ into another. This set gives a regular representation, so it contains all the i.r.'s. At first sight the Wannier function at $\mathbf{R} = \mathbf{0}$ seems to give all the $\mathbf{k}$'s the same phase. This, however, is not correct, for each Bloch function can be given any phase before the summation that defines a Wannier function is made. So it is possible to choose the phases in the summation over a star so that it transforms as any basis element of any i.r. of the point group. [There is a slight complication that for special choices of $\mathbf{k}$, e.g., $\mathbf{k}$ of the form $k_x = 0, k_y = 0, k_z = k$, the number of $\mathbf{k}$'s in the star is reduced so that it does not then contain all i.r.'s. A continuity argument based on the star of (ϵ, δ, k) as ϵ and δ tend to zero can be used to show that if a given i.r. is absent for $(0, 0, k)$ it is because a basis element of this i.r. for (ϵ, δ, k) becomes zero when the limit is taken.] The conclusion is that with the flexibility over the choice of n for the bands and the phases for the Bloch functions it is possible to arrange that each star in the summation that defines a Wannier function at $\mathbf{R} = \mathbf{0}$ gives a function that transforms as a specific basis element of a specific i.r. of the point group. The Wannier function therefore also has this property. The translational symmetry of the Wannier functions then shows that this property extends to all the cells. [An account of the symmetry properties of band states can be found in Cracknell (1975).]

The above result depends on the assumption that the lattice has just one atom per unit cell (a Bravais lattice). The assumption that the $\mathbf{k} = \mathbf{0}$ contribution at $\mathbf{R} = \mathbf{0}$ is p_x-like is a symmetry relation, which can be justified on the grounds that p_x is one of the basis elements of an i.r. of the cubic group. There are plenty of other functions which have this symmetry, so there is no guarantee that the Wannier function at $\mathbf{R} = \mathbf{0}$ is really like a hydrogenic p_x. And certainly there is no guarantee that it is as localized. To make it more p_x-like it would seem sensible to pay attention to the choice of the potential and, perhaps, choose it to be the superposition of many self-consistent atomic-like potentials, with each one centered on a different lattice point. Close to a particular nucleus the nuclear attraction can then be expected to dominate the effective potential, so the low-lying Bloch functions should be similar to free ion functions at that site. However, as the distance from the nucleus is increased the Bloch function can be expected to distort so that, near the next nucleus, it returns to the same free ion function with an altered phase. (This is also an assumption of the tight-binding model.) It seems likely that the Wannier functions, in a particular cell, will show similar atomic properties near the nucleus but that further away the only property that is retained is that associated with the symmetry of the lattice. Any degeneracies associated with a central potential are likely to be lifted in the sense that while xy-, yz-, and zx-like states will be symmetry related there will be no similarity with the $(x^2 - y^2)$- and $(3z^2 - r^2)$-like states. (This result may be contrasted with that implicit in crystal field theory, where the attempt is made to describe such distortions by combining states with the same l and therefore the same radial wave function.) The introduction of a periodic potential, which has wells near lattice sites, and the step to Wannier functions have the potential for providing an alternative and better way of describing ions in lattices.

To complete the step it is necessary to go on to lattices in which there are inequivalent ions. The periodic potential could then be constructed by superimposing the potentials of the set of ions, which will again lead to Bloch functions and associated Wannier functions. There is, though, a complication, for the definition of a Wannier function includes a summation over **k** vectors, so implying that the function will be localized to a cell rather than to specific sites within the cell. One possibility is to change the potential within the unit cell so that the various nuclei appear to occupy a selection of lattice points of a background lattice which does have one site per unit cell. This will allow the construction of Wannier functions at all the sites in the background lattice. A model of the electronic distribution in the actual unit cell can then be obtained by placing the electrons in a suitable selection of the Wannier functions, omitting those Wannier functions that relate to lattice points not occupied by nuclei. The objection that the resultant state may be very different from that which actually occurs can be met by the observation that the true eigenstates should be found from the second quantized form of the Hamiltonian, and therefore the choice of basis states used in defining the annihilation and creation operators is irrelevant.

Nevertheless, another procedure will be followed here. (On examples it appears to be rigorous; no general proof has been found.) The tight-binding model suggests that as the number of inequivalent sites in a unit cell increases the number of bands per given energy range will also increase. It can therefore be expected that the number of close-lying (in mean energies) Wannier functions in a given unit cell will also increase with the number of inequivalent ions in the

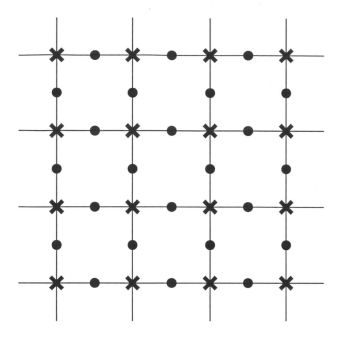

Figure 6.1. A square planar lattice structure that has two inequivalent ions in each unit cell.

unit cell. As these functions are mutually orthogonal and localized to the same cell, it seems not unlikely that they can be combined into other orthogonal sets to produce states associated with specific sites in the cell.

The reasoning will be developed in the context of the two-dimensional lattice shown in fig. 6.1, using fig. 6.2 which gives a labeling of the ionic arrangement in the vicinity of a specific lattice site. A Wannier function which is invariant under the local symmetry group at the lattice point labeled O will have the form

$$\alpha \sigma(O) + 1/2\beta[a(P) + b(Q) + c(R) + d(S)] + \cdots \qquad (6.21)$$

where $\sigma(O)$ denotes an invariant normalized function at O, $a(P)$ denotes a normalized function at site P, $b(Q)$ denotes the same function when it is rotated to site Q, and so on for $c(R)$ and $d(S)$, with α and β denoting amplitudes and $\cdots$ denoting the rest of the Wannier function. As this is assumed to be localized it can be expected that the part which is described by $\cdots$ will be small (see fig. 6.3, where the $\cdots$ part has been omitted). The Wannier function is therefore dominated by the parts in α and β and there are two possibilities, either one of α and β is much larger than the other or they are both of much the same

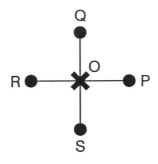

Figure 6.2. An ion at a corner of a unit cell and its nearest neighbors.

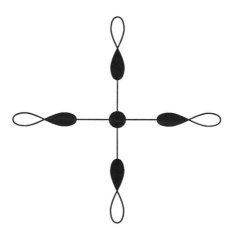

Figure 6.3. A schematic picture of a Wannier function centered on a corner of a unit cell. It is invariant under all rotations that leave the local symmetry unaltered.

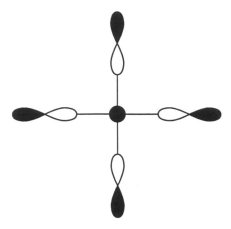

Figure 6.4. A Wannier function with the same symmetry but reversed signs for those of the neighbors.

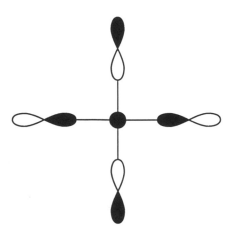

Figure 6.5. A Wannier function with a questionable symmetry. It illustrates that the symmetry of the neighbors is incompatible with that of the central ion. The function in its vicinity will therefore be completely absent or substantially modified.

magnitude. If α is much larger than β nothing more needs to be said, for the Wannier function is already well localized to O. If, however, α and β are of the same order another orthogonal function which has the same symmetry can be obtained by interchanging them and reversing one of their signs, followed by a slight adjustment in their magnitudes and some modification of the $\cdots$ omitted part (see fig. 6.4). It is not unreasonable to suppose that this new state can be identified with a Wannier function obtained from another band. The two Wannier functions can be combined to produce two orthogonal functions, one of which is localized to the central site and another which is delocalized to the sites P, Q, R, and S. The latter will therefore be rather similar to the function that would have been found had β been much large than α. Thus, depending on the magnitudes of α and β, which determine whether or not some initial recombinations should be made, it seems possible to produce Wannier-like functions that are either localized on the central site or delocalized onto the adjacent sites, with both functions having the same symmetry properties. The reasoning does not depend on the initial choice of symmetry. The fourfold rotational symmetry in the cluster shown in figure 6.2 directs attention to other possibilities, such as constructing four states of different symmetries which are delocalized to the neighbors. The initial choice has led to a delocalized orbital that can be represented by the form

$$a_1(P) + b_1(Q) + c_1(R) + d_1(S) \tag{6.22}$$

where the states used in (6.21) have been given the subscript 1 to denote that they were the states that were introduced with the first choice of symmetry. A second choice will give another orbital which is delocalized to the neighbors, one that has different phases for the different neighbors. It will have the form

$$a_2(P) + ib_2(Q) + i^2c_2(R) + i^3d_2(S) \tag{6.23}$$

where $i = \sqrt{1}$. With the form chosen for the periodic potential it can be expccted that $a_2(P)$ will he a function which is similar to, but not identical with, $a_1(P)$. Similarly, the third and fourth choices will produce

$$a_3(P) + (-1)b_3(Q) + (-1)^2c_3(R) + (-1)^3d_3(S)$$

and

$$a_4(P) + (-i)b_4(Q) + (-i)^2c_4(R) + (-i)^3d_4(S), \tag{6.24}$$

the first of which is represented in fig. 6.5. (The changes in the coefficients on moving from one state to the next are obtained by ringing the changes on the fourth roots of unity.) If the four functions are combined with phase differences that are also determined by the fourth roots, the result is the creation of four orthogonal functions, each of which is confined to one of the sites labeled P, Q, R, and S. (If the functions are simply added site P is selected.) The reasoning can be extended to more complicated lattices, leading to the conclusion that it is probably possible to obtain orthonormal functions that are localized to specific sites in each unit cell of any lattice, states that are also orthogonal to similar states in all the other unit cells. (The Wannier functions in any cell are orthogonal to those in any other cell. Forming linear combinations of those in any cell does not alter the orthogonality with states in any other cell.) It will be convenient to describe these states as Wannier states, though doing so requires an appropriate extension of the definition.

With the present example the $a_1(P)$ function has been given p_x-like symmetry, in which case the $b_1(Q)$ function, obtained by rotation, has p_y symmetry, the $c_1(R)$ function has $-p_x$ symmetry, and the $d_1(R)$ function has $-p_y$ symmetry. However, the translational symmetry associated with the standard definition of Wannier functions would have required the function at R to be identical with that at P, whereas it has the opposite sign. The ambiguity has arisen because although the $a_1(P)$, etc., can be regarded as Wannier functions they are not related by translations. This has happened because the i.r. associated with $a_1(P)$ is not one-dimensional. In considering rotational operations it is better to say that it is the i.r.'s of the Wannier functions or their projection operators which are translationally invariant.

A second quantized Hamiltonian formed using Wannier functions (multiplied by spin kets) looks quite different from that found using Bloch functions, for apart from its operators having different meanings the coefficients of the one- and two-electron matrix elements are also different. Nevertheless, the two second quantized forms will have exactly the same eigenvalues, though if some of these are degenerate it can easily happen that the eigenfunctions will look different. The question of whether to use a localized or a delocalized one-electron basis in setting up the second quantized Hamiltonian is therefore a question of which will be most tractable in perturbation theory, for it is unrealistic to assume that there is any basis that will lead to exact eigenvalues or eigenfunctions. With a diversity of perturbation methods and a diversity of orthonormal sets, the choice is not easily made.

For the rest of this book it will be assumed that second quantization has one feature which dictates its use, that it automatically incorporates the antisymmetric properties of electrons. To this will be added the requirement that if the system under consideration has any other symmetry then this too must be a feature of any unperturbed Hamiltonian.

§ 6.5 Occupation Probabilities

For any filling of a band it is possible to determine the probability that a Wannier function contains 0, 1, or 2 electrons by evaluating the expectation values of the appropriate number operators, which are best found by inspection.

For example, $a_{n+}a_{n+}^*a_{n-}a_{n-}^*$ applied to any one of $a_{n+}^*|v\rangle$, $a_{n-}^*a_{n+}^*|v\rangle$, and $a_{n-}^*|v\rangle$ gives zero. But applied to $|v\rangle$, the vacuum state which contains no electrons, it is equivalent to unity. So its expectation value, taken over any many-electron state, will give the probability that there are no electrons in the Wannier function that defines a_n and a_n^*. Similarly, the operator

$$[a_{n+}a_{n+}^*a_{n-}^*a_{n-} + a_{n-}a_{n-}^*a_{n+}^*a_{n+}]$$

gives the probability that one electron is present with either spin and $a_{n+}^*a_{n+}a_{n-}^*a_{n-}$ that two are present. These operator sequences can be changed into sums of products of creation and annihilation operators for band states, using

$$a_{n,\sigma}^* = \frac{1}{N^{3/2}} \sum_{\mathbf{k}} e^{-i\mathbf{k}\cdot\mathbf{R}_n} a_{\mathbf{k},\sigma}^* \qquad (6.25)$$

and its conjugate complex. For a many-electron state which is obtained by placing electrons in Bloch states the respective expectation values are

$$\frac{1}{N^3}\left(\sum_{\mathbf{k}_{e,+}}\right)\left(\sum_{\mathbf{k}_{e,-}}\right), \ \frac{1}{N^3}\left[\left(\sum_{\mathbf{k}_{o,+}}\right)\left(\sum_{\mathbf{k}_{e,-}}\right) + \left(\sum_{\mathbf{k}_{o,+}}\right)\left(\sum_{\mathbf{k}_{e,-}}\right)\right],$$

$$\frac{1}{N^3}\left(\sum_{\mathbf{k}_{o,+}}\right)\left(\sum_{\mathbf{k}_{o,-}}\right), \qquad (6.26)$$

where the $\mathbf{k}_{o,+}\mathbf{k}_{o,-}$ summations are to be taken over the occupied Bloch states and the $\mathbf{k}_{e,+}\mathbf{k}_{e,-}$ are to be taken over the unoccupied states. If the band is exactly half filled, which implies an average of one electron per site, the probabilities of finding 0, 1, and 2 electrons in a given Wannier function come out to be 1/4, 1/2, and 1/4, respectively. With the assumption that the Wannier functions are not all that different from atomic functions, these probabilities suggest that the sites will show three different ionicities, with the probabilities of ionicities 0 and 2 being equal to the probability that the ionicity is unity. When it is realized that this is for a band filling appropriate to a metallic conductor, these relative ionicities are surprising. (It seems to be predicting that in metallic sodium, which is usually assumed to have one electron per atom in a $(3s)$ conduction band, for a quarter of the time a given site will have no $(3s)$ electrons, for another quarter it will have two, and for half the time it will have one. That Na^+ and Na^- will be equally probable is easily understood, because if Na has lost an electron it has gone somewhere, which must be to another Na atom, to produce Na^-. Na^+ is a commonly occurring ionic state, in NaCl for example, whereas Na^- is an uncommon ion.)

§ 6.6 Model Hamiltonians

The presence of the two-electron interactions in the Hamiltonian of any system with more than one electron greatly complicates the problem of finding the eigenvalues and eigenfunctions. Nor is it much simplified by using second quantization. So many workers have chosen to work with model Hamiltonians in which second quantized forms are invented in the hope that they will be amenable to treatment and yet will contain aspects of the system which are expected to be important. For example, the ionicity problem mentioned above can possibly be dealt with by first ensuring that there is just a single conduction band, by having a summation in a model Hamiltonian of the form

$$\sum_{\mathbf{k},\sigma} \epsilon_{\mathbf{k}} a^*_{\mathbf{k},\sigma} a_{\mathbf{k},\sigma}, \tag{6.27}$$

where σ takes two values for the spin, with $\mathbf{k}$ being either a three-dimensional vector as for a real lattice or restricted to a one-dimensional lattice and denoted by k. Then to ensure that the band states are not completely occupied, when the description would be that of an insulator, it is necessary to specify that the number of electrons is less than $2N$, where N is the number of $\mathbf{k}$ vectors. Finally, to restrict the ionicities a term of the form

$$\sum_{n} U a^*_{n+} a_{n+} a^*_{n-} a_{n-}, \tag{6.28}$$

where the a^*_{n+} operators create Wannier functions at lattice sites, being themselves created from the operators for the Bloch functions, and U is a positive parameter (usually known as the Hubbard U), may be added to the model Hamiltonian. By letting U tend to infinity the possibility that a site will be doubly occupied is eliminated. Finding the eigenvalues and eigenstates of the model Hamiltonian is by no means a simple problem, even in this limit. However, if the number of electrons is chosen to be N the only possibility is that in each eigenstate the number of electrons at each site will be unity. The eigenvalue will be highly degenerate because of the twofold ambiguity of the spin orientation at each site and the model will describe an insulator. (No electron will be able to move from one site to another without introducing U.) The problem becomes more interesting if U is large rather than infinite, for although such excursions are then not out of question they are unlikely, and the system can be expected to remain insulating with the 2^N-fold degeneracy being partially lifted. (Total spin remains a good quantum number.) There is then the possibility of replacing the model Hamiltonian by an equivalent spin-Hamiltonian form, a spin of $1/2$ in each cell with Heisenberg-type interactions between the spins. When U is zero the Hamiltonian clearly describes a conductor, so another challenge is to determine the range of U values at which the model changes from one that describes a conductor to one that describes an insulator. The model can be elaborated in many other ways, such as by changing the number of electrons and/or introducing overlapping bands. An extensive literature has therefore grown up on Hubbard U models. Entry into the literature, particularly in the context of magnetism, can be obtained through Gautier (1982).

§ 6.7 References

Cracknell, A.P. 1975. *Group Theory in Solid-State Physics* (Taylor and Francis: London).

Gautier, F. 1982. *Magnetism of Metals and Alloys*, ed. M. Cyrot (North-Holland: Amsterdam), p. 1.

Wannier, G. 1937. *Phys. Rev.* 52:191.

Ziman, J.M. 1964. *Principles of the Theory of Solids* (University Press: Cambridge).

7

From Generic to Spin-Hamiltonian

§ 7.1 The Second Quantized Hamiltonian

With second quantization and Wannier functions, the study of a magnetic ion introduced as a substitutional impurity into an otherwise nonmagnetic host crystal can be reexamined. The procedure will be kept as similar as possible to that of crystal field theory, bearing in mind that the two cannot be identical because crystal field theory distinguishes between electrons.

For a substitutional magnetic impurity in a nonmagnetic crystal, the generic Hamiltonian which corresponds to that used in crystal field theory is that of a system of electrons moving in the fields of the nuclei and repelling one another by their Coulomb interactions. These are not the only interactions, but in making the comparison it is enough to be going on with. The Hamiltonian is taken in a second quantized form, using as basis a set of Wannier functions for the crystal before the magnetic ion is substituted. These functions will be assumed to transform as basis elements for i.r.'s of the point groups of the sites at which they are situated. It will also be assumed that each function can be put into 1:1 correspondence with one of the various one-electron wave functions that would be obtained in crystal field theory for a single electron placed at the same site.

§ 7.2 A (d^1) Example

To illustrate the procedure it is convenient to use a simple example. In Chapter 4, §4.2, a magnetic ion having one ($3d$) electron, surrounded by a regular octahedron of negative charges, was used as an introduction to crystal field theory. The same ion will again be considered, except that instead of there being six neighbors forming an octahedron there will now be a lattice of ions, the nearest neighbors of which form the octahedron. (There are plenty of crystals which have the necessary structure, MgO being one example.) It will be assumed that all the ions, other than the impurity, have, when isolated, a set of completely filled hydrogenic-like orbitals, with well-separated empty orbitals at higher energies. In the crystal the ionic arrangement will be much the same except that all the orbitals will be assumed to have been distorted into related orthonormal Wannier functions, with those that were previously occupied still being occupied and those that were empty remaining empty and well separated in energy from the filled orbitals. The electronic description is obtained by acting on a vacuum state which contains no electrons with a suitable product of creation operators, of which there are two, distinguished by a spin label, for each orbital state. The impurity is given a similar description except that as well as having a set of filled orbitals it will have a single electron in one of five orbitals, formed from distorting its $(3d)$ ionic states, each of which has two spin labels. These orbital states can be separated into two families, one containing three orbitals which transform like the xy, yz, and zx d orbitals (a T_2 i.r.) and another containing two orbitals which transform like $x^2 - y^2$ and $3z^2 - r^2$ (an E i.r.). Using z as axis of quantization the correspondences in symmetry are

$$|xy\rangle \equiv \frac{-i}{\sqrt{2}}[|2\rangle - |-2\rangle], \quad [|x^2 - y^2\rangle] \equiv \frac{1}{\sqrt{2}}[|2\rangle + |-2\rangle],$$

$$|yz\rangle \equiv \frac{i}{\sqrt{2}}[|1\rangle + |-1\rangle], \quad [|3z^2 - r^2\rangle] \equiv |0\rangle, \tag{7.1}$$

$$|zx\rangle \equiv \frac{-1}{\sqrt{2}}[|1\rangle - |-1\rangle],$$

where the states on the right of each equation are eigenstates of l_z, with $l = 2$, and all states are normalized. The Hamiltonian takes the form

$$H_{sq} = \sum_{(i,p),(j,q)\sigma} \langle i, p|h_1|j, q\rangle a^*_{i,p,\sigma} a_{j,q,\sigma} \tag{7.2}$$

$$+ \sum_{\substack{(i,p),(j,q),\\(k,r),(l,s),\\(\sigma_1,\sigma_2)}} \langle i, p : j, q|h_2|k, r : l, s\rangle a^*_{i,p,\sigma_1} a^*_{j,q,\sigma_2} a_{l,s,\sigma_1} a_{k,r,\sigma_2},$$

where $|i, p\rangle$ denotes the pth Wannier function at site i, h_1 is a one-electron operator given by

$$h_1 = \frac{\mathbf{p}^2}{2m} - \sum_J \frac{Z_J e^2}{|\mathbf{r} - \mathbf{R}_J|}, \tag{7.3}$$

where $\mathbf{R}_J$ is the position of the Jth nucleus, and h_2 is the two-electron operator, $e^2/2r_{12}$, which describes the Coulomb repulsions of the electrons. (The factor of $1/2$ is included because the summation in h_2 is over $i \neq j$.) The σ's are spin indices. Since h_1 and h_2 do not involve spin variables some of the matrix elements that would appear in the formal second quantized Hamiltonian have been dropped, being zero because of spin orthogonality. Where, in nonzero matrix elements, no spin index is required they have been omitted. The commutation rules for the creation and annihilation operators do not depend on the details of the initial Hamiltonian, so all its properties can be regarded as contained in the one- and two-electron matrix elements, which implies that no significance needs to be attached to the origin of the creation and annihilation operators. It will, in due course, be assumed that they are the annihilation and creation operators for the hydrogenic states of a system of well-separated ions. This does not imply that the second quantized Hamiltonian describes such a system; if it did it would have quite different matrix elements. Rather, use is being made of the feature that since it is only the matrix elements which matter the symbols that are used for the annihilation and creation operators can be changed to any other letter and interpreted as convenient. The only requirement is that the commutation rules between the various annihilation and creation operators, however they are denoted, remain unaltered.

In order to make the development as similar as possible to crystal field theory, several more steps are needed. Projection operators and their energies have to be defined for the various unperturbed manifolds, to define the unperturbed Hamiltonian of the Bloch perturbation expansion (§ 3.4). Then the perturbation expansion has to be examined with a view to finding an operator that gives the energy differences between the states into which the ground manifold splits under the perturbation. It will be shown that this can be written in a form similar to that used in crystal field theory.

The ground manifold consists of ten many-electron states, each of which will have pairs of electrons in the filled orbitals of the ions and one electron in a modified $(3d)$-like orbital on the magnetic ion, with either spin orientation. The filled shell arrangement is the same for all ten states. It can be constructed by operating on the vacuum state with an appropriate product of creation operators. If $b^*_{f\sigma}$ is an operator that creates an electron in a state of a filled shell with spin σ, which can take either of two values, to be denoted by $+$ and $-$, all that is necessary is to use a product of these operators, letting f range over all the filled orbitals and σ over the two spin values. The ten different states are then obtained by applying one of a further set of creation operators, $d^*_{n\sigma}$, where d is used as a reminder that a d-like state is being occupied, with n denoting that it is one of five possibilities. The precise meaning of n will depend on the local symmetry. In the present example the following identification will be made:

$$n = 1: \qquad |xy\rangle \equiv \frac{-i}{\sqrt{2}}[|2\rangle - |-2\rangle],$$

$$n = 2: \qquad |yz\rangle \equiv \frac{i}{\sqrt{2}}[|1\rangle + |-1\rangle],$$

$$n = 3: \qquad |zx\rangle \equiv \frac{-1}{\sqrt{2}}[|1\rangle - |-1\rangle], \qquad\qquad (7.4)$$

$$n = 4: \qquad |x^2 - y^2\rangle \equiv \frac{1}{\sqrt{2}}[|2\rangle + |-2\rangle],$$

$$n = 5: \qquad |3z^2 - r^2 > \equiv |0\rangle.$$

$n = 1$, 2, and 3 denote the states with T_2 symmetry and $n = 4$ and 5 the states with E symmetry. The projection operator for the ground manifold is then the sum of the projection operators of each of the ten many-electron wave functions. Many excited states exist, for any operator that moves one or more of the electrons into a state that is different from the ten in the ground manifold can be regarded as belonging to an excited manifold. The excited manifolds, which are needed in defining the unperturbed Hamiltonian, can be specified in terms of products of operators which act on states in the ground manifold.

Once the states in each manifold have been chosen, the generic Hamiltonian is used to determine the energies, the E_n's, which are to appear in the unperturbed Hamiltonian. They will be given a general definition for, in some cases, the expectation values of the generic Hamiltonian, taken over the states in a particular manifold, are not all the same. The E_n to be associated with a specific P_n is taken to be the mean expectation value of the generic Hamiltonian taken over the states in P_n. This ensures that each manifold is degenerate and that the unperturbed Hamiltonian incorporates parts of all the interactions in the generic Hamiltonian. Indeed, if the states chosen in the definition of a particular P_n had happened to be degenerate eigenstates of the Hamiltonian, then the E_n would have been their eigenvalue.

For the moment attention will be concentrated on states in the ground manifold, for they are the only ones needed in first-order perturbation theory, an expression in which the Hamiltonian is bordered by the projection operator of the ground manifold. This concentrates attention on operators that, acting on the states in the ground manifold, produce states in the same manifold.

In the one-electron part, h_1, of the second quantized Hamiltonian, each operator is a product of a starred and an unstarred operator, with the unstarred operator to the right of the starred one. If the action of the unstarred operator is to annihilate an electron in a state that is not already occupied it can be dropped, for it gives zero. If, however, it annihilates a state that is occupied then there are two possibilities, that the state is one in a fully occupied shell or it is a d state. In the former case the only nonzero action open to the starred operator to the left of the unstarred one is to restore the electron to the state from which it has been removed by the unstarred operator. The only pairs of this type which have a nonzero effect are therefore of the form $f^* f$, where both operators refer to the same filled state. As its overall effect on any state in the ground manifold is to leave it unaltered, it can be replaced by the unit operator. The associated matrix element can then be transferred into E_0, the unperturbed energy of the ground manifold. The only pairs that can contribute to the resolution of the degeneracy of P_0 are those in which the unstarred operator removes an electron from a d state and the starred one returns it, either to the same or to a different d state, without change of spin. (It has been assumed that h_1 has no spin dependence.)

However, before going further with the nonzero one-electron operators, it is better to examine the two-electron ones, for the two can be lumped together. If the effect of two unstarred operators to the right of two starred operators is to remove electrons from filled shells, the only operators that give nonzero contributions are those in which the two starred operators replace them. It is then instructive to use the commutation rules to show that they too contribute constant terms which can be moved into E_0. The technique is to move the starred operators to the right until they act on the filled-shell structure, when they give zero. On the way the commutation rules will have produced a remainder which is either 1 or -1. It is not necessary to go into the details, for the result will clearly be just a constant contribution to E_0. The analysis does, though, show a slight difference from the case of one-starred one-unstarred pairing, for the magnitudes and signs of the matrix elements depend on the order of the operators. The combination $f_1^* f_2^* f_2 f_1$, which is equivalent to unity (it vanishes if f_1 is the same as f_2), is multiplied by the matrix element $\langle f_1, f_2 | h_2 | f_1, f_2 \rangle$, whereas $f_1^* f_2^* f_1 f_2$, which can be rearranged into $-f_1^* f_2^* f_2 f_1$, is multiplied by $\langle f_1, f_2 | h_2 | f_2, f_1 \rangle$. The two matrix elements can be combined, with one having a sign that is the reverse of the other. Reading from right to left, the combination $f_1^* f_2^* f_2 f_1$ has first removed an electron from f_1 and then removed another from f_2. The creation operators have then restored both electrons, one into f_2 and the other into f_1 in the same order. The alternation in sign can be attributed to the difference in the initial sequence of removals. Each sequence can be replaced by the unit operator or by its negative, depending on the order in which the two electrons have been removed. The effect is as if each two-electron operator of this type has disappeared and left an associated matrix element which can be regarded as a contribution to E_0. (One matrix element will be of Coulomb type and the other will be of exchange type. The latter will be zero unless f_1 and f_2 have the same spin.)

With just one electron in the d shell the only other two-electron operators that need to be considered are those which remove one electron from a filled shell and the electron in the d shell, subsequently replacing them, one into the empty state in the filled shell and the other into the d shell, though not necessarily into the state from which it was removed. The removal of an electron from a filled shell followed by its replacement implies that in the sequence of two starred and two unstarred operators $f_{1\sigma}^*$ will occur to the left of $f_{1\sigma}$. The commutation rules can then be used to replaced the pair by 1 or -1. A $d_n^* d_m$ pair, such as is already likely to be present from h_1, will be left. The two can be combined and the overall effect will be to change the coefficient that multiplies a $d_n^* d_m$ pair. Its coefficient will have been "renormalized." Again the same operator can arise in different ways. Thus $d_{n+}^* f_{i+}^* d_{m+} f_{i+}$, which effectively moves an electron from one d orbital to another, has a multiplying matrix element of the form $\langle d_n, f_i | e^2/r_{12} | f_i, d_m \rangle$, whereas $d_{n+}^* f_{i+}^* f_{i+} d_{m+}$, which has the same set of operators but arranged in a different order, has a quite different multiplying matrix element, $\langle d_n, f_i | e^2/r_{12} | d_m, f_i \rangle$. When $n = m$, the first is of the Coulomb and the second is of the exchange form.

First-order perturbation theory has thus reduced the second quantized many-electron generic Hamiltonian to a one-electron form such as would be found for the second quantized Hamiltonian of a one-electron Hamiltonian that describes the motion of a single d electron. The link with the angular momentum form found with crystal field theory can now be established, by interpreting the annihilation and creation operators as operators acting on the d states of an isolated one-electron ion.

The d states provide a basis for a D_2 representation of the rotation group and so do the tensor angular momentum operators with $l = 2$. It is therefore possible to express any operator which connects just a single bra to a single ket, say $\langle m = 2 |$ and $| m = -1 \rangle$ for example, as a linear combination of the O_n^m. The required angular momentum operator is equivalent to a multiple of $l_+^3 [(l_z - 2)(l_z - 1)l_z(l_z + 2)$, where the string of $(l_z - r)$ factors on the right has been chosen to ensure that the operator has zero matrix elements for all other choices of the bra and ket. The annihilation and creation operators in the second quantized form derived using first-order perturbation theory appear in the form $a_n^* a_m$, where the operators were based on the states on the left-hand sides of the relations in (7.4). The reinterpretation of their meanings amounts to assuming they are based on the states on the right-hand side. So as every $a_n^* a_m$ pair can be reexpressed as a sum of tensor operators, it is now apparent that this holds for the whole of the second quantized form derived using first-order perturbation theory. The manipulations involved in doing so may, perhaps, be regarded as tedious. If so, it is fortunate that they are largely avoidable, for the end result can be inferred. At all stages the point group symmetry has been retained, so any angular momentum form must also have this symmetry. With cubic symmetry and $l = 2$ there are only two Hermitian forms in $\mathbf{l}, l_z, l_+$, and l_- which are invariant. The first is $(\mathbf{l} \cdot \mathbf{l})$, which can be replaced by $l(l + 1)$ and dropped as it corresponds to an overall displacement, and the second, using a fourfold axis for Oz, is the combination $O_4^0 + 5O_4^4$ [in the (A-B) notation]. Thus the form is exactly the same as that usually derived using crystal field theory. However, the coefficient which multiplies it has quite a different origin, for it is a combination of the matrix elements of h_1 and h_2 evaluated using Wannier functions. (While it has been convenient to use an example of an ion at a site of cubic symmetry it is not essential, for the reasoning is readily extended to any symmetry. In general, the effective operator will be a more complicated expression in $l = 2$ tensor operators.)

The picture that is emerging is broadly as follows. In forming a crystal all the orbitals of the ions will be distorted so that they are mutually orthogonal and form sets which are bases for the i.r.'s at their positions. On placing a magnetic ion at a lattice site of cubic symmetry, the d orbitals will be distorted into a family of three states having xy, yz, and zx symmetry and another having $x^2 - y^2$ and $3z^2 - r^2$ symmetry. An electron placed in any one of the first group will have a different energy from that which it would have were it to be in either state of the second group. For example, its kinetic energy will be altered, an effect which has no counterpart in crystal field theory. Another difference is the "electrostatic energy," for although the matrix elements of Coulombic form can be interpreted as describing crystal-field-like energies, there are no corresponding ways of interpreting the matrix elements of exchange type. They have no analogues in classical mechanics or in crystal field theory. So while the extra contributions may be small the physical picture of the origins of the so-called crystal field coefficients in the widely used tensor forms has been changed in a physically significant way.

To complete the investigation it would seem that a good deal more is required, an examination of all the higher orders in perturbation theory, though this is seldom done in standard crystal field theory unless the spin-dependent operators are present. Without them the determination of the general form given by the higher-order perturbation terms in the Bloch expansion and a second quantized Hamiltonian is not as complicated as it might seem. On looking at the numerators of the terms it can be seen that each is going to contain a product or a sum of products of many annihilation and creation operators. If one of the products is to give a nonzero contribution to the effective Hamiltonian, it will have to represent a sequence of excitations and deexcitations in which a state initially in the ground manifold eventually returns to the ground manifold. So if an operator f_σ occurs somewhere in the sequence, where f denotes a filled state in the ground configuration, then somewhere in the sequence a corresponding f_σ^* must occur to its left. The operators can be moved, using the commutation rules, until they are adjacent, when they can be replaced by 1 or -1, depending on the number of interchanges needed to bring the two together. Continuing in this way, the length of any sequence can be substantially reduced. Similarly, if e_σ appears somewhere, where e denotes an orbital which is empty in the ground configuration and which is not a d like orbital, then e_σ^* must have occurred to its right. Again, the pair can be brought together and replaced by 1 or -1. Such manipulations will eliminate most of the operators in any sequence, to leave, apart from constants, only those that either move the d electron to another state in the same shell or leave it where it is. (The constants that appear in orders higher than the first cannot be absorbed into the definition of E_0. They represent displacements of the manifold as a whole. It is only when the interest is in the splittings of the ground manifold that they can be dropped.) It can therefore be seen that all that happens in going to higher order in perturbation theory is that more operators of the same forms as those which have probably already occurred in first order will be produced. If so, they can be absorbed into the effective Hamiltonian of first order by renormalizing its parameters.

This reasoning demonstrates two important features of the combination of second quantization and the Bloch perturbation expansion: that it preserves all the symmetries at each order of perturbation theory and that the bordering P_0 operators have a very restricting effect on the form of the final effective Hamiltonian.

There is, though, one strange feature of the Bloch perturbation expansion, which has been ignored so far, that from its derivation it appears to give an effective operator which is non-Hermitian. The above analysis has assumed that the effective Hamiltonian is Hermitian, which is consistent with the observation that the first few orders in the perturbation expansion are Hermitian. It is not easy to determine the order in the expansion at which the Hermitian character disappears, so the assumption has been made that it is at a sufficiently high order that any non-Hermitian contributions to the effective Hamiltonian are negligible. There seems to be no experimental evidence that they exist.

§ 7.3 Several d electrons

With several d electrons a broadly similar treatment can be given, though a few changes are needed. The ground manifold is likely to contain many more states. For example, with just two electrons the number of states in the $(3d^2)$ configuration goes up to forty-five, and the number rapidly increases as the half-filled shell configuration is approached. From the point of view of writing down the second quantized Hamiltonian an increasing number of d electrons is no problem, for the form of the Hamiltonian is independent of the number. Nor is there any difficulty in describing the configurations and their projection operators. There is, though, a complication which shows up immediately, in first order of perturbation theory, which is the matter of simplifying an effective Hamiltonian when it contains operators that rearrange several d electrons in a configuration that is too large to be a basis for an i.r. of the rotation group. (This will inhibit the use of the Wigner-Eckart theorem. It is also an example where the expectation values of the Hamiltonian have several different values, so that E_0 is the mean energy of a configuration, not that of a term.) The same difficulty is, in principle, present in crystal field theory. There it is usually circumvented by assuming that the crystal fields act within the lowest term of the ground configuration, rather than within the whole configuration. It amounts to assuming that the crystal field splittings are small compared with the energy separations of adjacent terms. As the assumption seems to work well in crystal field theory, it seems not unreasonable to assume that it will also hold in the new way of treating the same ion when it is substitutional in a lattice. The P_0 operator should then be defined as a sum of the projection operators of the states produced by using those combinations of the creation operators that give the ionic $(d)^n$-like states spanning the L, S term of the ground configuration (along with the operators which create the electrons in the filled-shell Wannier states). The E_0 is then closer to that of a term. The physical requirement that the actual states are different from ionic-like states only shows up in the E_n's and the matrix elements that multiply the operators in the second quantized Hamiltonian.

In first order of perturbation theory the operators that remove electrons from filled states have to be accompanied by operators that replace them if they are to give nonzero contributions. As before, any $f_1^* f_1$ combination can be replaced by 1 or -1, so the only operators which merit attention are those that rearrange the electrons in the ground term. These will include two-starred two-unstarred operators as well as some one-starred one-unstarred operators of the one-electron example. As for the $(d)^1$ example, the bordering P_0's now span i.r.'s of the rotation group, so each nonzero operator can be rewritten as an expression in tensor operators in the L of the ground term. The first-order effective Hamiltonian can therefore be arranged to have the same tensor operator form as would be obtained using crystal field theory, except for one new feature. In crystal field theory the expansion of the crystal field potential is terminated at the $x^4 + y^4 + \cdots$ terms because all higher-order terms have zero matrix elements when evaluated between ionic wave functions that have $l = 2$. With L greater than 2 and underlying one-electron wave functions that do not have $l = 2$, there is no such restriction, though, depending on the value of L, there is a limit to the number of tensor operators (to $2L + 1$). In the ground terms of the $(3d)^n$ ions the only L values that occur are 0, 2, and 3 and it is only in $L = 3$ that there are operators equivalent to the sixth-order terms in the crystal field expansion. (This can be seen because some of the operators will include L_+^6 and L_-^6, which have matrix elements between $M = 3$ and $M = -3$. Any higher powers of L_+ are necessarily equivalent to zero.) With cubic symmetry the equivalent Hamiltonian should therefore contain $[O_6^0 - 21 O_6^4]$ along with $[O_4^0 + 5 O_4^4]$, using the (A-B) definitions. (It does not seem to have been needed in fitting experimental observations.) As these are the only operators that are invariant under the cubic group, no others will be introduced by going to higher order in perturbation theory. The factors which multiply them will not be the same as those occurring in crystal field theory.

§ 7.4 Spin-Dependent Interactions

So far, all spin-dependent interactions have been omitted from the generic Hamiltonian. In the case of a free ion the largest interaction involving the spin is the spin–own-orbit interaction, though there are others, such as the spin–other-orbit and the spin-spin interactions. For the moment, the spin–own-orbit interactions will be enough to go on with. Choosing it presents a problem, for in second quantization all electrons are treated on an equal footing, which implies that in a periodic system any electron is equally likely to be at any equivalent point in the lattice. The form of the spin-orbit interacton is not obvious (Yafet, 1963). About the best that can be done is to choose it so that it is compatible with the symmetry of the system which, for a cubic lattice, amounts to writing it as $\mathbf{u} \cdot \mathbf{s}$, summed over all the electrons, where $\mathbf{u}$ is a vector function of the position and momentum variables of an electron that has the periodicity of the lattice. (In a free ion the spin-orbit coupling contains $\mathbf{l}$, which is odd under the operation of time reversal. It can be expected that $\mathbf{u}$ will also have this property, in which case it is likely to contain momentum operators.) With a substitutional impurity the periodicity is lost, so only the point group and the time-odd symmetries have to be satisfied. It is therefore convenient to assume that the only large spin-orbit interactions are those associated with the d-like Wannier states of the impurity. These add

$$\sum_{\substack{i,\sigma_1 \\ j,\sigma_2}} \langle i, \sigma_1 | \mathbf{u} \cdot \mathbf{s} | j, \sigma_2 \rangle d^*_{i,\sigma_1} d_{j,\sigma_2}, \tag{7.5}$$

to the second quantized Hamiltonian, where i and j label the orbital d states. As before, any operator pairs that, in the perturbation expansion, involve moving electrons out of and then back into states in closed shells can be replaced by constants. So once more the closed shells can be ignored in determining the form of the effective Hamiltonian. Each combination of annihilation and creation operators can be regarded as acting first on the orbital parts of the basis states and then on the spin parts, and since L and S define distinct i.r.'s of the rotation group each orbital part can be replaced by a sum of tensor operators in L and each spin part by a sum of tensor operator in S. When combined they must then produce an expression which is invariant under the symmetry operations of the point group. In cubic symmetry an obvious invariant combination is $\mathbf{L} \cdot \mathbf{S}$, though there are a number of others, depending on the magnitudes of L and S. The cosine form is the only one that arises with cubic symmetry and first order of perturbation theory. (With tetragonal symmetry there are two obviously related invariant forms, $L_z S_z$ and $L_x S_x + L_y S_y$, when the fourfold axis is taken as Oz. With trigonal symmetry there is a similar expression, when Oz is taken along the threefold axis.) The step of going to higher order in perturbation theory will not be explored in detail, for enough examples of how to manipulate pairs of creation and annihilation operators have been given to make it clear that doing so is possible. The main problem is coping with the large number of matrix elements and the ways in which they need to be combined. The general result is that they renormalize the coefficients of various expressions, most of which will already be present in first order.

Examples that assume that the symmetry is cubic are somewhat unrepresentative because in most crystals the symmetry is lower. The basis states will have the symmetry properties of whatever symmetry remains. It is often so low that the i.r.'s are one-dimensional, in which case each P_n will be the projection operator of a single state. But if the level patterns can be regarded as due to a slight distortion from cubic symmetry, it is probable that the E_n's will form closely spaced groups of levels, such as, in crystal field theory or in Bethe's group theoretical treatment, can be regarded as having arisen from the resolution of an i.r. of the cubic group. There is then the option of choosing the P_n's and the E_n's as above or of lumping the P_n's of the close-lying levels together into a composite projection operator with an energy that is the mean energy of the set. In many cases, the latter procedure is to be preferred because in higher orders of perturbation theory it eliminates the energy denominators arising from closely spaced energy levels. The alternative, which regards each level as an orbital singlet, results in the angular momentum being quenched in first order, so there is nothing to lift any spin degeneracy. It is then essential to go on to at least second order to examine the effects of spin-dependent interactions. With the former choice the spin-dependent couplings between states in closely spaced groups will show up in first order, and often the couplings to remote states will be sufficiently small that they are negligible. Then an adequate effective Hamiltonian will emerge from a first-order theory. The only complication is that of deciding how it is best expressed. For example, in cubic symmetry there may be an orbital state which has threefold degeneracy. If it transforms as T_1 all the parts of the operator combinations that relate to orbital states can be expressed in terms of tensor operators which have $L = 1$, for its three components also span a T_1 i.r. But what can be done if the orbital states transform as T_2, a more common case? Doing so with $L = 1$ is clearly possible because with L_z, L_+, and L_- it is possible to construct operators that, using L_z as axis of quantization, will represent any matrix that has zeros everywhere except at one position. So it is possible to represent a T_2 by an $L = 1$, but only by proceeding rather carefully—and avoiding group theory. In a similar way a split E doublet can be represented by an $S = 1/2$, though to avoid confusion with spin it is often expressed as a $T = 1/2$. The effective Hamiltonians for these cases, often still described as spin-Hamiltonians, will then contain operators from two families of angular momentum operators, L or T to encompass the orbital parts and S for the spin parts. Having set them up using specific choices for the axes of quantization, this is no longer necessary. Any convenient axes can be used.

For an isolated singlet orbital state (effective $L = 0$) with $S = 1/2$ there will be no splitting of the spin degeneracy, in any order of perturbation theory, by Kramers' theorem, unless an external magnetic field is present. For any larger spin nonvanishing operators can be constructed from the components of S, and only those that are compatible with the point group symmetry will occur. They will have the same forms, combinations of tensor operators in S, as the zero-field splitting operators in typical spin Hamiltonians, such as are derived by crystal field theory, by inspired guesswork, or by fitting experimental results.

Including spin-spin and spin–other-orbit interaction will only renormalize the coefficients of these operators. With spin equal to an integer plus one-half, Kramers' theorem shows that all energy levels are of even degeneracy. It is a feature of the tensor operator forms that this result emerges automatically.

The application of an external magnetic field eliminates most of the symmetry of any Hamiltonian. This does not prevent the same projection operators and their energies being used to define the unperturbed Hamiltonian. It simply means that at the final stage, when the effective second quantized Hamiltonian is being expressed in angular momentum form, some of its operators will have a lower symmetry than that of the point group. The final form may contain operators that are not found using crystal field theory. If so, this it is probably because crystal field theory has not been carried to high enough order. It does not contain the equivalent of the bordering projection operators of the Bloch theory, so the restrictions which these impose may only emerge as the perturbation theory is developed. The commonly used version is only valid to second order. This has usually been enough—except for S-state ions.

Turning to the rare earth ions, the second quantized theory can be made to look similar to the crystal field treatment by decomposing the ground configuration even further, into submanifolds of different J's. After that the course follows that already describe for L manifolds, when it gives effective Hamiltonians in which the tensor operators are expressions in J, J_+, J_-, and J_z. In crystal field theory the expressions in powers of x, y, and z are terminated at x^6, etc., because all higher powers give zero matrix elements between states with $l = 3$. The highest power that J_+ then has, in the crystal field effective Hamiltonian, is J_+^6. However, the highest J value in the ground multiplets of the rare earths occurs with $(4f)^{10}$ Ho^{3+}, when it is 8. There are therefore states in this multiplet that differ by a ΔM_J of 16 and in the Bloch treatment the only requirement that restricts J_+ is that an operator must be nonvanishing. It can therefore be expected that operators that contain J_+ up to J_+^{16} may appear in the effective Hamiltonian for Ho^{3+}, using the Bloch scheme, with only slightly lower powers for some of the other rare earth ions. There does not seem to be any indication that such high powers have been needed in fitting experimental results. It seems unlikely that the distortions of the $(4f)$ orbitals in a crystalline environment will be sufficiently large that they will produce observable operators of this form.

While there have been many tests which show that spin-Hamiltonians are good effective Hamiltonians for describing the low-lying levels of magnetic impurity ions in insulating crystals, it does seem that most of the iron group examples have been on ions for which the orbital degeneracy has been substantially lifted. The above analysis shows that there are good reasons for expecting that such effective Hamiltonians should exist not only for these examples but also for ions in which there are groups of low-lying orbital states. Also, although this has not been stressed, the same sort of reasoning should be applicable to excited groups of levels. There is then the challenge of relating the level pattern of a low-lying group to that of a higher-lying one.

Going back to the beginning of the analysis provides a reminder that it has been based on Wannier functions that have been constructed from the Bloch functions of a periodic lattice for which no specific choice of periodic potential has been made. They were therefore to some extent arbitrary. Even so, if they provide a complete orthonormal set this is of no consequence, for the properties of the second quantized form of the generic Hamiltonian are independent of the choice. This cannot be claimed, however, for the parameters that multiply the angular momentum tensor operators in the spin-Hamiltonians when they have been derived by first-order or even second-order perturbation theory. Yet the impression may have been given that these are being compared with quantities obtained by experiment. This is where the infinite order of the Bloch perturbation scheme is relevant, for it is presumably only when the perturbation treatment is taken to high enough order that the parameters can be expected to converge to the experimental values. They should then become invariant to changes in the underlying orthonormal basis set. It is for this reason that more emphasis has been placed on the form of the tensor operators in the spin-Hamiltonian than on the values of the parameters which multiply them and which are renormalized as the perturbation procedure continues.

§ 7.5 Displaced Wannier Functions

So far it has been convenient to assume that when a magnetic ion substitutes for a nonmagnetic ion in a crystal the lattice structure is unaltered, for this has meant that the Wannier functions of the regular lattice have been available to describe the configurations, etc., of the whole crystal. It is less clear that it is a realistic assumption, for it seems more likely that the structure in the vicinity of an impurity will be distorted. In attempting to include such a possibility it would seem necessary to have a set of orthonormal functions which might be described, assuming they exist, as the Wannier functions of a distorted crystal or displaced Wannier functions. The theory of this section should therefore be regarded as more speculative than most of what has gone before, for it is an attempt to define and then use such a set. It begins with a standard result, that if an operator, $\exp(iS)$, where S is Hermitian, is applied to every member of an orthonormal set of functions, the outcome is a new set of orthonormal functions. The question of finding a set of displaced Wannier functions can therefore be reduced to that of finding a suitable form for S.

At this point it is instructive to seek an answer to a simpler question: is it possible to find an S that displaces every Wannier function by the same amount, so that it appears as if the whole lattice has moved bodily? The answer is well known, for the operator with $S = \delta \cdot \mathbf{p}/\hbar$, where δ is the displacement and $\mathbf{p}$ is the momentum operator for an electron, has just this property. This is readily seen in a one-dimensional example, for

$$[\exp(i\lambda p)]x = \sum_{n=0}^{} \left[\frac{(i\lambda p)^n}{n!} \right] x \tag{7.6}$$

and

$$p^n x = x p^n - i\hbar n p^{n-1}, \tag{7.7}$$

so that

$$[\exp(i\lambda p)]x = (x + \hbar\lambda)\exp(i\lambda p) \tag{7.8}$$

or

$$\exp(i\lambda p)x \exp(-i\lambda p) = x + \hbar\lambda. \tag{7.9}$$

The above S, with $\delta \cdot \mathbf{p}$ replacing λp, where δ and $\mathbf{p}$ are three-dimensional, is the transformation required to give each function the same displacement, δ. This, however, is not what is wanted, for the requirement is that every Wannier function at a given site should have a displacement that is special to its site. It does though suggest that a better choice might be to set

$$S(\delta) = \frac{-1}{(2\hbar)} \sum_{J} [(\delta_J \cdot \mathbf{p}) P_J + P_J (\mathbf{p} \cdot \delta_J)], \tag{7.10}$$

where δ_J is the displacement at site J and P_J is the projection operator of all the Wannier functions at site J. (The expression has been symmetrized, to be Hermitian, there being no reason to suppose that $\mathbf{p}$ and P_J commute.) It has the property that if all the δ_J are put equal to the same value, δ, then δ can be taken outside the summation to leave a summation over all the projection operators, which gives the unit operator, and so reduces the form to that which gives a uniform displacement of all the Wannier functions.

Armed with this new operator the second quantized Hamiltonian can be set up in the new basis, not forgetting that if the nuclei are displaced the one-electron terms that describe the Coulomb interactions between the electrons and the nuclei will have to be modified. Each a^* creation operator in the previous form will need to be replaced by $\exp(iS)a^* \exp(-iS)$ and there will also have to be a change in the definition of the vacuum state. However, as far as the second quantized Hamiltonian is concerned, it will make no difference whether the operators are given a different notation or whether they are left unchanged, for the operators after the transformation have exactly the same commutation properties as those which their "parent" operators had before the transformation. The significant changes will be in the matrix elements in the second quantized Hamiltonian. For example, a matrix element such as $\langle i, p | h_1 | j, q \rangle$ in (7.2) becomes

$$\langle i, p | \exp(iS) h_1 \exp(-iS) | j, q \rangle$$

and $\langle i, p : j, q | h_2 | k, r : l, s \rangle$ becomes

$$\langle i, p : j, q | \exp[i(S_1 + S_2)] h_2 [\exp[-i(S_1 + S_2)]] | k, r : l, s \rangle,$$

where S_1 and S_2 are, respectively, the expressions for S when it is written in the variables of electrons 1 and 2. h_1 should also be replaced by $h_1(\delta)$. So what has happened is that the creation and annihilation operators have stayed the same as those for the undistorted lattice, and it is the first quantized Hamiltonian which has been changed, so resulting in a change in the matrix elements of the second quantized Hamiltonian.

On the assumption that the displacements are small the matrix elements can be expanded in powers of the δ's. The leading term, which is independent of the displacements, is of no present interest for it plays no part in determining whether the equilibrium structure of the lattice has been altered by substituting the magnetic impurity. These are primarily determined by the operators that are linear and quadratic in the δ's. (In one-dimensional classical mechanics a potential energy of the form $\alpha q + \beta q^2$, where q is a displacement measured from a particular origin and β is positive, implies that $q = 0$ is not the equilibrium condition. The potential energy is a minimum when $q = -\alpha/2\beta$. In three dimensions a similar analysis can be made, though the expressions that determine the equilibrium position are more complicated.) In quantum mechanics the quantities corresponding to the α's and β's are operators, and an equilibrium position can hardly be the ratio of two operators. In many cases it is assumed that the electronic ground state is a singlet. The α's and β's are then replaced by their expectation values taken over the ground state, and the equilibrium position is found as in classical mechanics. It should, though, be emphasized that in quantum mechanics a nucleus cannot be at rest at an equilibrium position, for this is incompatible with the uncertainty principle. It is therefore best to regard an equilibrium position as the point about which oscillations occur. If the ground state is not a singlet the question which is then raised is that of whether or not there is a concept of "equilibrium position."

To go more deeply into the question of what happens when there is a degenerate electronic ground state, it is usual to introduce generalized coordinates for the displacements, a step that is most easily demonstrated with a model. In a simple cubic lattice an impurity, denoted by 0, will have six nearest neighbors, forming a regular octahedron (see fig. 2.1), which can be numbered from 1 to 6. The central ion and each neighbor are given displacements and the components of δ_n at site n are denoted (X_n, Y_n, Z_n). On applying a symmetry operation of the cubic group the twenty-one quantities $X_0, X_1, X_2, \ldots Z_6$ are permuted, so they form a basis for a reducible representation of the group. Its decomposition into i.r.'s shows that there will be twenty-one linear combinations, new "displacements," Q_n, each of which is a basis element of an i.r. (A similar procedure can be used for the star of displacements of more remote ions in the lattice.)

Once the terms linear in the δ's have been rewritten in terms of the Q_n's for the whole lattice the second quantized operator expressions which they multiply can be arranged into subsets of the form $\sum Q_n V_n$, which are invariant under the cubic point group operations. (If, for example, the Q_n's for $n = 10$ and 11 transform as E then V_{10} and V_{11} must also transform as E to make $Q_{10}V_{10} + Q_{11}V_{11}$ an invariant.) Each V_n can then be replaced by a reduced matrix element times an orbital tensor operator. If the local symmetry gives an orbital ground state that is a singlet, most of the angular momentum operators give zero when averaged over the ground state. The only nonzero contributions are numerical multiples of invariant Q's, from which it follows that the symmetry is preserved in the equilibrium lattice. (Ni^{2+} and Cr^{3+} are examples of ions that have ground orbital singlets.) If the ground state is not an orbital singlet then it is much more difficult to determine what happens.

§ 7.6 The Jahn-Teller Effect

When the ground orbital level is degenerate, the terms linear in the Q's become invariant sums of products of tensor angular momentum operators and Q's that transform in related ways. For example, in a cubic lattice the operators $(L_x L_y + L_y L_x)$, $(L_y L_z + L_z L_y)$, and $(L_z L_x + L_x L_z)$ transform as the T_2 i.r., so if they appear in an expression that is linear in the δ's they must occur in the form

$$Q_{xy}(L_x L_y + L_y L_x) + Q_{yz}(L_y L_z + L_z L_y) + Q_{zx}(L_z L_x + L_x L_z),$$

where Q_{xy}, Q_{yz}, and Q_{zx} transform, respectively, in the same way as $(L_x L_y + L_y L_x)$, $(L_y L_z + L_z L_y)$, and $(L_z L_x + L_x L_z)$. Taking the above form as an example, the operators can be replaced by expectation values, but only if a specific choice is made of one of the states in the degenerate ground manifold. After the linear terms have been eliminated by a change of origin (moving to a new equilibrium) the lattice has a lower symmetry and the orbital degeneracy is lifted. With a different choice of state, staying within the ground manifold, a similar conclusion is reached, except that the equilibrium position has moved and the ground state is different from that found with the previous equilibrium position, which raises the question of how the ground orbital state should be chosen. There are a few examples, using crystal field theory, for which definite answers can be given. One is for an ion in an octahedral environment which has, as its lowest level, an orbital triplet of T_2 symmetry that is coupled only to distortions of E-type symmetry (Stevens, 1969). For each of the three orthogonal orbital ground states there is a different equilibrium distortion, all of which have the same energy. This shows that there is no resolution of the threefold degeneracy and that the "vibronic" states span a T_2 i.r. ("vibronic" will be defined shortly).

When this sort of problem was first encountered, by Jahn and Teller (1937), the conclusion was reached that there would usually be a spontaneous lowering of the local symmetry that would lift the degeneracy, a phenomenon which came to be known as the "static Jahn-Teller effect." Shortly after, Van Vleck (1939) examined the stability of a threefold degenerate ground level of a magnetic ion in an octahedral crystal field, with the assumption that the displacements were quantum mechanical variables and that with each component of displacement there would be an associated momentum variable. This change produced a Hamiltonian that was invariant under the operations of the cubic group, so leading to the conclusion that the states in any degenerate energy level would form a basis for a representation of the group. Further, with his particular ground state he knew that ignoring the dynamics of the lattice the orbital states of the ion transformed as a three-dimensional i.r. He therefore reasoned that on switching on the coupling to harmonic oscillators that were in their ground states, a singlet vibrational state, the symmetry would be unchanged. The degeneracy would therefore not be lifted but the states within it might well be much more complicated in structure, for they were states in the Hilbert space of orbital and lattice vibrational states. Such states are now described as "vibronic." This conclusion was at variance with the symmetry-lowering prediction of Jahn and Teller.

Since then a continuing interest in the Jahn-Teller effect has arisen, for it occurs in a wide variety of physical systems, in most of which the theoretical treatment presents problems. So there is now an extensive literature devoted to it, as well as to the controversy it has generated. Examples are not confined to magnetic ions; for those that are, crystal field theory in its spin-Hamiltonian form is invariably used, to describe how the magnetic ions are affected by displacements of neighboring ions. There is probably little to be gained at present by examining how it would be changed if electrons were no longer distinguished, for it is progress with the present models that is most needed.

Strange as it may seem, although Van Vleck's analysis was published in 1939 there was one very relevant question that was not answered until many years later (see Ham, 1969). It can be posed as follows. If the Jahn-Teller effect does not resolve the degeneracy, how can its presence be recognized? The answer amounted to showing that various quantities, and particularly the orbital angular momentum, which appears in crystal field theory in the spin-orbit coupling and in the Zeeman interaction, could be expected to show a much reduced contribution, by as much as an order of magnitude, to some of the parameters in spin-Hamiltonians. Substantial reductions have been observed (see A-B), and attributed to what is known as the dynamic Jahn-Teller effect.

There is a complication, though, for one of the best ways to observe the consequences of the dynamic J-T effect is to study impurity magnetic ions in otherwise nonmagnetic hosts by EPR. This, however, is not possible, for crystals do not contain just a single magnetic impurity. The most that can be done is to study crystals having a low concentration of impurities, magnetic and nonmagnetic, in the hope that the local symmetry at the J-T ion has not been sufficiently lowered that it has removed the orbital degeneracy and so eliminated the effect that is being sought. It is therefore possible to claim, in experiments in which the degeneracy is observed to be lifted, that this is due to a "static Jahn-Teller effect," whereas it may be due to the presence of nonmagnetic impurities or of other J-T ions. Apart from this difficulty, there is another, that any experimental method that is set up to observe the effect is itself likely to impose a lowering of the symmetry. It has therefore been particularly important to find examples where even in the presence of symmetry-lowering impurities the orbital angular momentum parameters are much reduced (Bates, 1978).

Another manifestation of the Jahn-Teller effect occurs with crystals that are fully concentrated in J-T magnetic ions, ions at lattice points where the symmetry is so high that it can be expected that they would have degenerate ground states had they been present as isolated ions. At high enough temperatures it is usually found that the magnetic centers appear to be almost independent of one another, which can be explained by first assuming that J-T distortions are occurring and then arguing that, if they are, the splittings will be so small compared with kT that all the states resulting from the splitting will be thermally populated, and the lowering in magnetic energy due to the splitting will be smaller than the elastic energy needed to create it, so there will be no J-T distortion. If the temperature is lowered it will eventually reach a value below which there will be an overall reduction in energy, and then the total energy can be decreased by distortions. While the form that these will take is open to dispute, there is evidence that in some cases the whole lattice undergoes a structural transformation at a specific temperature, which can be expected to lift the degeneracy of the ground state of each ion. The phenomenon is known as the "cooperative Jahn-Teller effect."

Probably the most quoted example of a cooperative Jahn-Teller phase transition is that of the Cu^{2+} ion in octahedral coordination. This ion would be expected to have an orbital doublet lowest, with spin 1/2, and yet in many crystals the Cu^{2+} ions are found to have a singlet orbital state lowest and four neighboring nonmagnetic ions in a square planar array, with two further neighbors displaced away from the positions that they would have in a regular octahedron. So the observed symmetry is "explained" as due to a cooperative J-T transition, from a regular octahedral coordination at a much higher temperature to a tetragonal one at some lower temperature—for which there is generally no experimental evidence! The most convincing experimental evidence for the phenomenon is found with some of the rare earth vanadates (Gehring and Gehring, 1975).

§ 7.7 Spin-Lattice Relaxation

Most of the information about the low-lying states of isolated magnetic ions has come from EPR, with a lesser amount coming from inelastic neutron scattering and optical spectroscopy, all of which are forms of spectroscopy. They rely on the absorption of quanta that match energy level differences. There is, though, only a net absorption if there is a population difference between the initial and final states, such as occurs in thermal equilibrium. The absorption process lifts ions from low-lying states to higher ones, so disturbing the populations. It follows that unless there is some mechanism whereby excited ions can return to the initial state without giving energy back to the stimulating source, the absorption will eventually cease. This situation, which is known as saturation, can almost be reached in studying a number of ions by EPR. In most EPR experiments, saturation is avoided by reducing the microwave power. Otherwise, the absorption on resonance is much reduced and the line shapes are distorted, making it difficult to resolve closely spaced lines. The phenomenon is, though, of interest in its own right, so it has been investigated. In the spectroscopy of free ions with low intensity e.m. radiation saturation does not occur because the ions in excited states can return to initial states by the emission of electromagnetic radiation, either into modes of the electromagnetic field that exclude the mode being used for the excitation or into the mode that is being used. In the latter process the energy emitted is returned to the source, whereas in the former it is lost from the source. It is this loss, described as spontaneous emission, which is responsible for the net absorption of energy. In EPR the e.m. frequencies are so low that spontaneous emission is an ineffective mechanism for returning ions from excited states back to lower-lying states in comparison with the possibility of spontaneous and even stimulated emission into the vibrational modes of the crystal lattice. So the study of what is known as spin-lattice relaxation has arisen as an offshoot of EPR.

The lattice vibrations can be regarded as lowering the local symmetry in a time-dependent way, so producing time-dependent splittings of the orbital levels. The theory, which is almost entirely due to Van Vleck (see Stevens, 1967), is based on crystal field theory. Recasting it so that the electrons are not distinguished is now possible, though no major differences are to be expected.

§ 7.8 References

Bates, C.A. 1978. *Phys. Rep.* 35:187.

Gehring, G.A., and Gehring, K.A. 1975. *Rep. Prog. Phys.* 38:1.

Ham, F. 1969. *Jahn-Teller Effects in Electron Paramagnetic Resonance Spectra, Electron Paramagnetic Resonance* (Plenum: New York).

Jahn, H.A., and Teller, E. 1937. *Proc. Roy. Soc.* A161:220.

Stevens, K.W.H. 1967. *Rep. Prog. Phys.* 30:189.

Stevens, K.W.H. 1969. *J. Phys. C: Solid State Phys.* 2:1934.

Van Vleck, J.H. 1939. *J. Chem. Phys.* 7:72.

Yafet, Y. 1963. In *Solid State Physics*, ed. F. Seitz and D. Turnbull (Academic Press: New York), Vol. 14, p. 2.

§ 7.9 Further Reading

The article "Jahn-Teller Effects in Paramagnetic Crystals" (Bates, C.A. 1978. *Phys. Rep.* 35:188) provides an introduction to the theory of J-T ions as impurities in ionic crystals. It contains a number of spin-Hamiltonians in which triplets and doublets in cubic symmetry are represented, respectively by $L = 1$ and $T = 1/2$. Another introduction can be found in "Co-operative Jahn-Teller Effects" (Gehring, G.A., and Gehring, K.A. 1975. *Rep. Prog. Phys.* 38:1). There is also the book by R. Englman (1972) *The Jahn-Teller Effect in Molecules and Crystals* (Wiley: London).

The topic is actively developing and for some time international symposiums have been held about every two years. The conference reports can be found referenced under "Jahn-Teller."

8

The Interactions between Ions

§ 8.1 General Considerations

It has long been known that there are interactions between magnetic ions, and a natural extension of the spin-Hamiltonian concept is to assume that given two magnetic ions in a crystal each would have its own spin-Hamiltonian and there would be interactions between them, expressible in terms of their effective spins. Some authors went further and suggested that the Hamiltonian for the pair could be split into two single-ion crystal-field-like Hamiltonians, one for each ion, with a perturbation which represented the interaction between them, the implication being that a simple extension of crystal field theory would give the expected spin-Hamiltonian for each ion together with their interactions. As far as your author is aware this program has never been carried out, though at first sight it seems a reasonable proposition (see Stevens, 1976, §2.1).

The effective or spin-Hamiltonians for many magnetic ions present in very low concentrations ($\sim$ parts per million) in numerous crystals have been determined by electron paramagnetic resonance. At these low concentrations the ions are so well separated that the interactions between them are extremely small and the experiments throw little light on them. To obtain more information it is necessary to increase the concentrations. Several problems then arise. Many different types of pairs can be expected, so although EPR is an excellent technique for examining individual ions, it can be difficult to disentangle the properties of a specific pair from a superposition of results on many different pairs. A further complication is that in many cases the dominant interaction has an antiferromagnetic character which couples the members of a pair so that the ground state is nonmagnetic. There are then no resonances at low temperatures. Going to higher temperatures may result in higher-lying levels being populated, but even so the resonances may be too weak because of the relatively small population differences between Zeeman-split levels. There may also be complications due to zero-field splittings. (A pair has an even number of electrons so the overall spin is an integer.) Nevertheless, if the resonances can be observed they probably give more detailed information about the interactions within the pairs than can be obtained by any other technique.

An important extension has been to study crystals in which the structure is such that all the magnetic ions occur as closely spaced pairs. There are a fair number of these, though when the study first began, in the early 1950s [see (A-B), p. 506], copper acetate was about the only known example. It was the study of this compound by EPR that confirmed the existence of antiferromagnetic interactions. The individual ions in such pairs are unlikely to be on crystallographically identical sites, so it may not be possible to regard them as resulting from the substitution of magnetic ions in an otherwise undistorted crystal (Barry et al., 1981; Hirst and Ray, 1982). That antiferromagnetic interactions occur in many magnetic crystals became much more firmly established following the development of neutron scattering techniques, which also showed that the strongest interactions can be those between magnetic ions that are separated by nonmagnetic ions and so are not nearest neighbors (Shull et al., 1951).

The absence of the above information did not inhibit an interest in the theory of Heisenberg interactions, for the algebraic problem presented by an effective Hamiltonian

$$H = \sum_{\langle i,j \rangle} \mathbf{s}_i \cdot \mathbf{s}_j, \tag{8.1}$$

where $\langle i, j \rangle$ denotes a summation over nearest-neighbor spins, does not depend on the sign of the exchange interaction. [For this reason the magnitude and sign have been omitted from (8.1). For the theoretician it is just a multiplicative factor of the expression on the right-hand side. For the experimentalist the sign is very important, for a change inverts the whole energy level pattern.] Such a Hamiltonian presents a challenging and frustrating theoretical problem; it possesses a lot of symmetry, invariance under lattice rotations and translations (assuming periodic boundary conditions), invariance under spin rotations, and commutation of each term with the total spin. Yet it is almost intractable. Even today not much is known about its eigenvalues and eigenfunctions in any lattice other than that of a linear chain. Using approximate methods, it was soon shown that it could account for the presence of the Δ's in the Curie-Weiss law, but by far the major theoretical challenge is to show that it predicts that a phase transition will occur as the temperature is lowered.

In this book the view will be taken that the present-day interest in effective Hamiltonians of the above form is a mathematical exercise, for it is unlikely to represent any actual physical system, for two reasons. The first is that any effective Hamiltonian is likely to be more complicated in its form and have lower symmetry, and the second is that the experimental properties of the antiferromagnets in their low-temperature phases cannot emerge from Hamiltonians of the above form. (There is nothing in the Hamiltonian that will determine the direction of alignment of the spins relative to the crystallographic directions.) To substantiate the first reason it is necessary to have a theory for the interactions. It is doubtful whether this can be satisfactorily developed using crystal field theory, for it inevitably seems to distinguish between some of the electrons. The theory of Chapter 7 has no such problem and can readily be extended to pairs, which is the main topic of this chapter. It will be shown that the inclusion of spin-orbit coupling complicates the interionic interactions and leads to the expectation that the ground state of the Heisenberg problem bears little resemblance to the ground state of the effective Hamiltonian obtained when the spin-orbit interactions are included.

The original theory for the coupling between magnetic ions amounted to assuming that they were effectively free spins of $1/2$ with ferromagnetic Heisenberg-like exchange interactions between nearest-neighbor ions, with much weaker magnetic dipolar interactions between all ions, interactions which decrease as $1/r^3$, where r is the separation of the two ions. The realization that the observed interactions were more usually of antiferromagnetic character caused a major rethink, and it was P. W. Anderson (1963), in the 1950s, who laid the foundations for a proper understanding of the origin of antiferromagnetism.

§ 8.2 Two d^1 Ions

It is convenient to begin by assuming that two identical magnetic ions have been substituted for two nonmagnetic ions, in a nonmagnetic crystal. For the present it will also be assumed that the periodicity of the basic lattice is unaltered and that its Wannier functions are known. With cubic symmetry and d^1 ions the ground orbital state will be degenerate, which is inconvenient in an introductory account, so it will be assumed that the local symmetry is sufficiently reduced that each ion has a singlet orbital ground state. (Even in cubic symmetry the presence of the other magnetic ion will reduce the local symmetry, but it is preferable not to rely on this to produce the singlets.) The Bloch form of perturbation theory will again be used, with the generic Hamiltonian modified to allow for the presence of two magnetic ions. This in turn leads to modifications of the projection operators. For the ground level P_0 will be the sum of four many-electron projection operators, each of which will be a projection operator of states that have a common closed-shell structure and a common arrangement of d-like orbitals. Where they differ is in the four spin arrangements: $++$, $+-$, $-+$, and $--$ associated with the two singly occupied d-orbitals. There will be many excited manifolds, with associated projection operators, some of which will be similar to those for a single impurity, excitations of electrons from filled to empty orbitals on the same ion, and others in which electrons are moved from one ion to another. In particular, there will be states in which the d electron on one ion has been moved into the d shell of the other ion, states that have no counterpart in crystal field theory nor in the analysis in the previous chapter. The important point is, though, that whichever state is of interest, it can be obtained by operating on the vacuum by an appropriate set of Wannier creation operators.

§ 8.3 First-Order Perturbation Theory for $(d)^1 - (d)^1$

The generic Hamiltonian contains many different interactions. The usual procedure is to retain only those that are thought to be of interest, knowing that if it is subsequently realized that some interaction that has been neglected ought to have been included then it is not difficult to incorporate it in first order of perturbation theory. With the Hamiltonian bordered by P_0 operators the number of operators that give nonzero contributions is very restricted, and many of these reduce to constants which can be moved into E_0. As with the single-impurity problem, the first step is to "eliminate" the electrons in closed shells, so reducing P_0 to an operator that acts in a four-dimensional Hilbert space. The only operators that remain are those which would be there for a pair of electrons in d-like Wannier states. They divide into two sets, one containing operators of the form d^*d, where both operators refer to the same site, and the other containing forms that can be arranged to be of the type $(d^*d)(d^*d)$, where the first $(\ldots)$ refers to one site and the second to the other. [Each nonvanishing operator in P_0 conserves the number of electrons at each site, and a useful property is that every (d^*d) pair for one site commutes with every pair for the other site.] Those of the first type have appeared in the treatment of the single-site problem, except that, in the matrix elements associated with the Coulomb interactions of the electrons with the nuclear charges, there will now be a change, because of the altered nuclear charge at the site of the other magnetic ion. The symmetry at each magnetic ion is not, therefore, the same as it would be for an isolated impurity. For a single d electron and a singlet orbital ground state, the single-site pairs, in first order, are of no consequence, because by Kramers' theorem the spin doublet cannot be split, except by an external magnetic field. Their contribution to the effective Hamiltonian is that of a free spin of $1/2$, however they have arisen. (There are also contributions, to E_0, from the changed number of electrons at the site of the other impurity.) The spin-orbit interactions are also of one-electron type. The bordering P_0 operators eliminate any processes that might transfer electrons from one site to the other, so again the only operators left are similar to those for isolated $(d)^1$ ions, where it is known that their contribution is zero due to orbital quenching.

The two-starred two-unstarred operators which conserve the number of electrons at each site are much more interesting. They can only have come from the two-electron terms in the Hamiltonian, which focuses interest on two forms:

$$\langle a_{\sigma_1}, b_{\sigma_2} \left| \frac{e^2}{r_{12}} \right| a_{\sigma_1}, b_{\sigma_2} \rangle a_{\sigma_1}^* b_{\sigma_2}^* b_{\sigma_2} a_{\sigma_1} \tag{8.2}$$

and

$$\langle a_{\sigma_1}, b_{\sigma_2} \left| \frac{e^2}{r_{12}} \right| b_{\sigma_1}, a_{\sigma_2} \rangle a_{\sigma_1}^* b_{\sigma_2}^* a_{\sigma_2} b_{\sigma_1}, \tag{8.3}$$

where it has been convenient to change the notation so that a denotes the d orbital at one site, A, and b denotes the d orbital at the other site, B. The spin indices have now been given explicitly in the matrix elements, for while this is not really necessary it shows how the accompanying operators are related to their matrix elements. In both expressions, the operators can be rearranged, using the commutation rules, to give a product of (a^*a) and (b^*b) forms, expressions which commute. Taking (8.2) first, the value of the matrix element does not depend on the choice made for σ_1 and σ_2. Summing over the four spin possibilities results in the matrix element being multiplied into

$$(a_+^* a_+ + a_-^* a_-)(b_+^* b_+ + b_-^* b_-), \tag{8.4}$$

which can be replaced by 1, for the first factor is the number operator for site A and the second is the number operator for site B. So the whole of (8.2), like so many of the other operators in the second quantized form, can be moved into E_0. The operator part of (8.3) when summed over all spin orientations becomes

$$[a_+^* b_+^* a_+ b_+ + a_+^* b_-^* a_- b_+ + a_-^* b_+^* a_+ b_- + a_-^* b_-^* a_- b_-]. \tag{8.5}$$

If $(1/2 + s_z)_1 (1/2 + s_z)_2$ is written in its second quantized form it gives $a_+^* a_+ b_+^* b_+$ which is the negative of the first operator in (8.5). Similarly, the next two can be recognised as negatives of the second quantized form of $(s_+)_1(s_-)_2 + (s_-)_1(s_+)_2$, and so on. So (8.3) can be written as

$$-\langle a, b \left| \frac{e^2}{r_{12}} \right| b, a \rangle (\frac{1}{2} + 2\mathbf{s}_1 \cdot \mathbf{s}_2).$$

The matrix element is of exchange type and positive and the spin-dependent part in the second factor is isotropic in spin, so the expression reduces to a Heisenberg ferromagnetic exchange interaction between two localized spins of $1/2$. [Since

$$2\mathbf{s}_1 \cdot \mathbf{s}_2 = (\mathbf{s}_1 + \mathbf{s}_2) \cdot (\mathbf{s}_1 + \mathbf{s}_2) - \mathbf{s}_1 \cdot \mathbf{s}_1 - \mathbf{s}_2 \cdot \mathbf{s}_2) = S(S+1) - 3/2$$

the minimum occurs when $S = 1$.] What was a second quantized form in annihilation and creation operators has been replaced by a first quantized operator, which is that of two spins of $1/2$ coupled by a ferromagnetic exchange interaction of cosine form. Further, the spin components of one commute with the spin components of the other. (This is usually taken for granted.)

An alternative way of simplifying (8.5) is to use the commutation rules to move the operators for a given site so that they become adjacent. Thus $a_+^* b_+^* a_+ b_+$ becomes $-(a_+^* a_+)(b_+^* b_+)$, a product of two commuting quantities. $(a_+^* a_+)$ can then be replaced by $(1/2 + s_z)_A$ and $(b_+^* b_+)$ by $(1/2 + s_z)_B$, which removes the need to regard $a_+^* b_+^* a_+ b_+$ as a two-electron operator.

§ 8.4 Higher-Order Perturbation Theory for $d^1 - d^1$

The first-order theory has shown that by combining the Bloch perturbation scheme with a second-quantization version of the Hamiltonian it is possible, in what is the simplest problem involving two magnetic ions, to obtain an effective Hamiltonian which has the same form as that usually obtained by adding an isotropic exchange interaction to the spin Hamiltonians of two separated ions obtained using crystal field theory. To agree with most observations, the sign chosen in crystal field theory is, though, the opposite of that which has come from first-order perturbation theory. From the work of Anderson it seems likely that there will be terms of similar forms in higher orders and that some of these will be large enough to reverse the overall sign of the interaction. That interactions of the necessary form will occur is fairly certain, for all the higher-order expansion terms are bordered by the same P_0. The numerator of a typical higher-order term in the Bloch expansion is a product of the form $P_0 V P_1 V P_2 V \ldots P_0$, which, apart from the bordering P_0 operators, is an alternation of the perturbation with projection operators of excited manifolds. When such a product is examined in detail it is seen to consist of a sum of "processes" each of which has a numerator that is a product of matrix elements found in the second quantized Hamiltonian and a product of annihilation and creation operators. A large number of these processes can be dropped because they do not reduce to operators that couple states in P_0. From the examples already given, it is easy to see what operator forms will be left.

Many of these processes will have pairs of operators that move electrons out of and then, at some stage, back into an orbital that is occupied in P_0. Such pairs can be replaced by 1 or -1, to leave, in many cases, operators which have already appeared in first order and which can be incorporated by renormalizing the first-order parameters. What is of particular interest is to see whether there are any processes that produce isotropic exchange interactions with an antiferromagnetic sign and whether there are others that are not invariant in spin space. (The latter should couple the spin alignments to the crystallographic axes.)

On comparing (8.2), which does not contribute to the exchange interaction, with (8.3), which does, an obvious difference is that (8.3) is a process that reverses oppositely directed spins of electrons at two sites, whereas (8.2) leaves the spin directions unaltered. It is therefore of interest to examine the numerator of a second-order process that also reverses the spins. An example is

$$P_0 \langle a|h_1|b\rangle a^*_+ b_+ P_n \langle b|h_1|a\rangle b^*_- a_- P_0, \qquad (8.6)$$

Reading from the right the first two operators move the d electron from site A and place it, with the same spin, in the d orbital at site B. This is only possible if the electron initially at A has negative spin and the electron at B has positive spin. There are now two electrons in the d orbital at B and no d electron at A. So the intermediate state is an excited state. The second two operators move an electron with positive spin from B to A, so restoring the initial charge arrangement with reversed spins. The intermediate state belongs to a P_m other than P_0. There is no need to identify it, for unless it is identical with the P_n in (8.6) the whole expression vanishes, and if it is identical the presence of P_n is unnecessary. This illustrates a useful feature, that the intermediate P_n operators can be dropped when particular processes are being examined. Using the commutation rules, (8.6) reduces to

$$- |\langle a|h_1|b\rangle|^2 a_+^* b_-^* b_+ a_-. \tag{8.7}$$

There will be a similar process in which all the operators appear with reversed spins. When the two contributions are added the expression is the second quantized version of

$$-|\langle a|h_1|b\rangle|^2 [(s_+)_A (s_-)_B + (s_-)_A (s_+)_B],$$

which is clearly not invariant in spin space. As there is no obvious reason why h_1 should produce such a result, there is the implication that there are further processes that will restore the rotational symmetry. They are likely to be processes that move the electrons in the same way without inducing the spin reorientations. An obvious choice is

$$|\langle a|h_1|b\rangle|^2 b_+^* a_+ a_+^* b_+$$

as one member of a pair, with the similar expression in which all the spins are reversed as the other. Together they are the second quantized version of

$$-|\langle a|h_1|b\rangle|^2 [1/2 + 2(s_z)_A (s_z)_B].$$

Along with a constant they provide the missing interaction, so eliminating the anisotropy in the spin-spin interaction. As each process is divided by the same difference of unperturbed energies, a negative expression, this contribution to the isotropic exchange interaction has the sign required for antiferromagnetism.

The conclusion that an antiferromagnetic interaction can arise from h_1 when it is used twice might seem to be in conflict with another argument that suggests that no term of the above form can possibly appear. If all two-electron interactions are dropped from the generic Hamiltonian, the part which is left has a one-electron spin-independent form and, apart from having to conform to the exclusion principle, the electrons will be independent of one another. There is then no possibility that the energy of a pair of electrons in different orbital states can depend on their relative spin orientations. The resolution of the difficulty lies in a property which is implicit in the Bloch form of perturbation theory, that the intermediate state in which an electron has been moved from A to B is an excited state. If the Hamiltonian had contained only one-electron operators this state would have had the same energy as the initial and final states and it would not have belonged to an excited P_n. So an important point has been illustrated, that in choosing the E_n's to form the unperturbed Hamiltonian, the full Hamiltonian should be used, not just a part of it. Obtaining an antiferromagnetic spin coupling from two uses of h_1 has also used h_2, in the energy denominator. [This feature can be relevant when comparisons are made between different perturbation schemes, for the Bloch expansion used here is not the only one: see, for example, Brandow (1977), §6.5.2 and Jefferson (1989).] In some the antiferromagnetic exchange interactions only appear in much higher orders. It may also be noted that as h_1 contains the kinetic energy an isotropic exchange is not necessarily entirely due to Coulomb interactions.

The above example of how an antiferromagnetic contribution to the exchange constant can arise has not examined the role that might be played by a nonmagnetic ion between the two magnetic ions. It is therefore of interest to consider using h_2 to transfer two electrons with opposite spins, one from A and the other from B, into an empty orbital on an intervening ion, and then to return the two electrons to their original orbitals but with reversed spins. A process that does this is

$$\langle a_+, b_- \left| \frac{e^2}{r_{12}} \right| e_+, e_- \rangle a_+^* b_-^* e_- e_+ \langle e_+, e_- \left| \frac{e^2}{r_{12}} \right| b_+, a_- \rangle e_+^* e_-^* a_- b_+, \quad (8.8)$$

where e^* is the operator that creates an electron in an empty state of the intervening ion. It can be rewritten as

$$- \left| \langle a, b \left| \frac{e^2}{r_{12}} \right| e, e \rangle \right|^2 a_+^* a_- b_-^* b_+, \quad (8.9)$$

an expression which is similar to that found in (8.7). However, unlike the example that produced (8.7), there is no similar process in which the initial state is one in which the two spins have the same spin alignments, for it would not then be possible to excite them into the same orbital while preserving the spin orientations. (8.9) and the process in which all the spins are initially reversed will be the second quantized version of an expression containing $(s_+)_1 (s_-)_2 + (s_-)_1 (s_+)_2$, which is an anisotropic spin-spin interaction. With no spin-dependent interactions in the Hamiltonian there is no way in which such anisotropy can arise, so it is clear that there must be some other processes that will restore the isotropy in spin. The two processes that move oppositely directed spins into the same empty excited orbital and then return them with no spin reversals are the ones required. The result is another isotropic spin-spin interaction of antiferromagnetic character.

There are many other processes that contribute to the isotropic exchange interaction, but for most of these it is not possible to determine the sign of the contribution without specifically evaluating matrix elements. The above examples are therefore of interest, because they show that there are processes that give opposite signs, which makes it possible to conclude that the experimental discovery of a sign that differs from that emerging from the Heisenberg-Dirac theory is not necessarily inexplicable. Indeed, it seems probable that the first-order ferromagnetic contributions to the exchange are often negligible compared with those from higher orders, showing that no valid conclusions can be drawn about actual systems from the H-D theory.

§ 8.5 Anisotropic Exchange Interactions

So far the analysis has used only the spin-independent terms in the Hamiltonian, and Heisenberg-like exchange interactions have emerged, as is to be expected. The inclusion of dipolar interactions between the magnet moments associated with the spins, such as are to be expected to be present in a classical lattice of magnetic moments, might well introduce anisotropy into the exchange interactions, as might the spin–own-orbit interactions, which are known to lead to anisotropy in the magnetic moments of ions at noncubic lattice sites. For an ion with only one d electron which is at a site of low enough symmetry that an orbital singlet is lowest, the quenching of the orbital moment results in the spin-orbit coupling giving zero in first order of perturbation theory. Any interactions from it therefore have to be sought in higher orders.

A simple second-order process, which is much like one already described, begins with the d electron at A having a minus spin and the d electron at B having a plus spin. This time, though, the spin-orbit interaction is used to move the minus electron at A into the ground orbital at B, so that it is doubly occupied and then to move the plus spin at B back to A, so effecting a mutual spin reversal. The sequence

$$\langle a_+|\,\text{s-o}\,|b_+\rangle a_+^* b_+ \cdot b_-^* a_- \langle b_-|\,\text{s-o}\,|a_-\rangle \tag{8.10}$$

uses only the part of the spin-orbit interaction, $u_z s_z$, that leaves the spin unaltered, yet it produces a spin reversal. When combined with a similar process in which the initial spin indices are reversed, the equivalent expression is

$$-\tfrac{1}{4}\langle a|u_z|b\rangle\langle b|u_z|a\rangle[(s_+)_A(s_-)_B + (s_-)_A(s_+)_B]. \tag{8.11}$$

Beginning with the same spin arrangement, another process is to transfer the minus electron at A to B and then transfer it back to A. This process gives

$$\langle a_-|\,\text{s-o}\,|b_-\rangle a_-^* b_- \cdot b_-^* a_- \langle b_-|\,\text{s-o}\,|a_-\rangle, \tag{8.12}$$

which, when combined with the similar process in which all the spins are reversed and with (8.11) gives, apart from a constant

$$\tfrac{1}{4}|\langle a|u_z|b\rangle|^2[(s_+)_1(s_-)_2 + (s_-)_1(s_+)_2 - 2(s_z)_1(s_z)_2], \tag{8.13}$$

a spin operator that is not invariant in the spin space. (If the s_z factors are omitted from the spin-orbit interaction the example becomes identical with the one in which a spin-independent h_1 is used, showing that it is the presence of the s_z's which has converted an isotropic interaction into an anisotropic one.) There are two related processes, obtained by reversing the order of the one-electron operators in (8.10) and (8.12). They can be included by doubling (8.13). There are no other second-order processes that will introduce $|\langle a|u_z|b\rangle|^2$ and, with the symmetry being low, there is no reason to suppose that $|\langle a|u_x|b\rangle|^2$, $|\langle a|u_y|b\rangle|^2$, and $|\langle a|u_z|b\rangle|^2$ will be algebraically related. The conclusion must therefore be that the spin-orbit coupling will lead to spin-spin interactions that are anisotropic in spin space.

While it has been convenient to use just the $u_z s_z$ parts of the spin-orbit interaction to demonstrate how spin-spin anisotropy can arise in second order of perturbation theory, a full discussion should also use $u_+ s_-$ and $u_- s_+$ and the rest of the perturbation and go to higher orders, all of which suggests that the final expression for the effective interaction will be quite complicated. In fact the bordering P_0 operators severely limit its form. To second order and omitting the negative energy denominator, a full investigation of the above processes gives

$$
\begin{aligned}
\tfrac{1}{2} \quad & |\langle a|u_z|b\rangle|^2 \left[\frac{1}{2} + s_A \cdot s_B - 3(s_z)_A (s_z)_B\right] \\
- \quad & \frac{1}{2}[\langle a|u_z|b\rangle\langle b|u_+|a\rangle + \langle b|u_z|a\rangle\langle a|u_+|b\rangle][(s_z)_A(s_-)_B + (s_-)_A(s_z)_B] \\
+ \quad & \frac{1}{4}[|\langle a|u_+|b\rangle|^2 + |\langle a|u_-|b\rangle|^2]\left[\frac{1}{2} + 2(s_z)_A(s_z)_B\right] \\
- \quad & \frac{1}{2}[\langle a|u_-|b\rangle\langle b|u_-|a\rangle(s_+)_A(s_+)_B \\
+ \quad & \langle a|u_+|b\rangle\langle b|u_+|a\rangle(s_-)_A(s_-)_B],
\end{aligned}
\tag{8.14}
$$

which is not isotropic in spin and looks more complicated than it really is, for its general form can be simplified by remembering that there is another constraint, that there should be no preferred directions for Ox, Oy, and Oz. The consequences of this symmetry lead to a general form for the effective interaction.

Showing that (8.14) conforms to this general form will be left to the reader as an exercise. The reasoning uses four different basis sets for the D_1 i.r. of the rotation group. Two of these use the components of the two spins and the other two are $\{\langle a|u_x|b\rangle, \langle a|u_y|b\rangle, \text{ and } \langle a|u_z|b\rangle\}$, and $\{\langle b|u_x|a\rangle, \langle b|u_y|a\rangle, \text{ and } \langle b|u_z|a\rangle\}$. (A rotation of axes replaces a given component of $\mathbf{u}$ by a linear combination of its components, leaving the bras and kets unchanged.) The relation $D_1 \times D_1 = D_0 + D_1 + D_2$ shows that by using the spin components an invariant D_0 combination can be formed, which is immediately recognizable as

$s_A \cdot s_B$. It can be written as $(s_A s_B)^0$ to indicate that the components of the spins have been combined to form a D_0 i.r. Similarly, the D_1 combination is given by the components of the vector product $s_A \wedge s_B$. It can be written as $(s_A s_B)^1$. The D_2 set is not quite so obvious and for the moment it will not be given. Similar D_0 and D_1 sets can be constructed using the matrix elements. Invariant expressions that involve all four basis sets are obtained from the scalar products of a D_0 formed using the spins with a D_0 formed from the matrix elements, from a D_1 formed from the spins with a D_1 from the matrix elements, and so on. The first of these is $s_A \cdot s_B$, or $(s_A s_B)^0$, with a scalar multiplicative factor composed of matrix elements. The second will be slightly more complicated: it can hardly be other than an expression which has the form $d \cdot s_A \wedge s_B$, where d is a vector in which all its components are expressions in matrix elements. It can therefore be written as $(dd)^1 \cdot (s_A s_B)^1$. The third invariant can also be deduced by inspection, using a well-established formula, the mutual energy of two dipoles:

$$\frac{m_1 \cdot m_2}{|r_{12}|^3} - 3\frac{(m_1 \cdot r_{12})(m_2 \cdot r_{12})}{|r_{12}|^5}. \tag{8.15}$$

That it is invariant under rotations of axes is clearly evident, as also is the fact that its form differs from the two invariants found so far. All that is therefore necessary to obtain the third invariant is to replace r by a unit vector t and multiply the whole expression by a scalar, to give

$$K[s_A \cdot s_B - 3(s_A \cdot t)(s_B \cdot t)], \tag{8.16}$$

where K and t are expressions in matrix elements. It can be written as a multiple of $(s_A s_B)^2 \cdot (tt)^2$. The $s_A \cdot s_B$ form is the familiar Heisenberg exchange interaction, often referred to as isotropic exchange. The second, $d \cdot s_A \wedge s_B$, is known as antisymmetric exchange.[That it might occur was suggested by Stevens (1953), and brought into prominence as an explanation of the canting of magnetic moments in some weakly ferromagnetic crystals by Moriya and Dzialoshinski, see Moriya (1963).] It is not present in (8.14) and indeed it is only present when the pair symmetry is particularly low. The third one, the dipole-dipole form, was introduced phenomenologically by Van Vleck (1937), in connection with anisotropy in cubic crystals. It is known as anisotropic exchange. It is probably sensible to assume that it is always present in the interaction between two spins. It may, of course, be very small, but it is interesting because it relates the interaction to the symmetry of the pair, so introducing a physical feature that is absent with the Heisenberg form, a coupling that relates spin alignments to crystallographic directions.

The group theoretical reasoning is independent of the order of the perturbation theory, so it shows that going to higher order can only result in renormalizations of the coefficients of the three rotationally invariant interactions. (It should be stressed that the rotations are rotations of axes that leave the ions in unchanged positions. Such rotations should be distinguished from rotations that change one pair into another.) The analysis has also confirmed a widely held assumption, that the effective spin operators at site A commute with the effective spin operators at site B. This arises because the states used in defining the P_0 projection operator conserve the number of electrons at each site, which means that in each expression in the effective Hamiltonian the number of creation operators that refer to a specific site is accompanied by an equal number of annihilation operators for that site. It is this structure that results in the spin operators at different sites commuting. The basic particles are electrons, which satisfy Fermi-Dirac statistics, yet the effective "particles," entities at different sites that possess spins, are described by commuting operators.

In the original theories of the paramagnetism of ionic crystals the moments were regarded as independent entities which could be treated using Boltzmann statistics and the same concept continued to be used when quantum theory took over, even though the electrons were known to obey Fermi-Dirac statistics. The above analysis shows that the continuity can be justified provided that an effective Hamiltonian of spin-Hamiltonian form exists. (A similar assumption is also used in some of the theories used to describe magnetic impurities in conducting hosts.)

It has been convenient to use two $(d)^1$ ions to demonstrate that interactions arise quite naturally when the second quantized method for isolated ions is extended to pairs. There are plenty of more complicated pair systems, of which the next simplest would be pairs of identical ions with several d electrons. Provided that they have nondegenerate orbital ground states the corresponding analysis, taken to second order, can hardly do more than replace the **s** operators for the A and B sites by **S** operators, for the degeneracy of the ground manifold is determined by the spins. On going to higher order more complicated interactions can be expected, for the $D_1 \times D_1$ decomposition used for the spins of $1/2$ depended on the result that any products of the components of a single spin are reducible to a scalar plus an expression linear in the components of total spin. As **S** is increased the number of independent expressions that can be made with its spin components is equal to $4S(S + 1)$. [A matrix representation of any component has $(2S + 1)^2$ positions. The number of independent expressions is one less because it can be assumed that the sum of the diagonal elements is zero.] Another difference is that the effective Hamiltonian for a pair will probably contain operators that describe the zero-field splittings of both ions. When either or both ions have orbital degeneracy the effective pair interaction will be even more complicated, but whether there is any value, at present, in extending the theory to such systems is questionable, for the experimental information is sparse.

There is still much to learn about exchange interactions, particularly in the context of rare earth ions, where it is expected that they will be smaller in magnitude than those found with iron group ions because the magnetic electrons are deeper in the ions. For the same reason, the role of the spin-orbit interaction is likely to be increased relative to that of the "crystal fields," so the theory should be based on a ground manifold described by $\mathbf{J}$'s, the total angular momentum operators. Since orbital and spin motion are then tightly coupled the exchange interactions are unlikely to be dominated by the isotropic form, in contrast with the iron group ions [though see Donni et al. (1988)].

§ 8.6 Comments and Further Reading

A more detailed account of pair theory is given in Stevens (1976). Many compounds are now known to have closely spaced pairs of magnetic ions. The use of EPR has been supplemented by optical spectroscopy. Transitions are induced between states in the lowest-lying manifold and states in higher-lying manifolds. The spectrum tends to entangle the structure of the ground manifold with that of excited manifolds. It is therefore best to choose the excited manifold carefully, so that the two can be separated (see Maxwell and Turner, 1991).

§ 8.7 References

Anderson, P.W. 1963. In *Magnetism*, ed. E.T. Rado and K. Suhl (Academic Press: New York), Vol. 1, p. 25.

Barry, K.R. 1981. *J. Phys. C: Solid State Phys.* 14:1281.

Brandow, B.H. 1977. *Adv. Phys.* 26:651.

Donni, A., et al. 1988. *J. Phys.* Colloq. (France) 49:C8-513.

Hirst, L.L., and Ray, T. 1982. *Proc. Roy. Soc.* A384:191.

Jefferson, J.H. 1989. *J. Phys. C: Solid State Phys.* 1:1621.

Maxwell, K.J., and Turner, R.J. 1991. *J. Phys. Chem. Sol.* 52:691.

Moriya, T. 1963. In *Magnetism*, ed. E.T. Rado and K. Suhl (Academic Press: New York), Vol. 1, p. 85.

Shull, C.G., et al. 1951. *Phys. Rev.* 83:333.

Stevens, K.W.H. 1953. *Rev. Mod. Phys.* 25:166.

Stevens, K.W.H. 1976. *Phys. Rep.* 24c:1.

Weiss, P. 1914. *Ann. Phys.* (Paris), 9^e s., t, 1:134.

Van Vleck, J.H. 1937. *Phys. Rev.* 52:1178.

9

Cooperative Systems

§ 9.1 Preliminary Remarks

The theory of pair interactions described in the previous chapter can be extended to fully concentrated magnetic crystals, when it produces an effective Hamiltonian in which each ion is described by a spin-Hamiltonian and all the ions are coupled to one another by exchangelike spin-spin interactions. This form is then usually simplified by assuming that the interactions between closely spaced ions dominate. It is difficult to claim that such a picture is supported by experimental observations, for unfortunately there are no techniques which give anything like the amount of detailed information about the energy levels of large clusters of magnetic ions that EPR does about isolated magnetic ions. The basic experimental information needed to check the physical picture in detail is simply not available at present. What is known is that in most fully concentrated magnetic crystals there is a phase change as the temperature is reduced, to what is known as a cooperative magnetic phase, and that at even lower temperatures it is possible to excite wavelike excitations which have dispersion diagrams which can be determined experimentally. [A wave traveling in a direction $\mathbf{k}$ has an angular frequency, ω, that depends on $\mathbf{k}$. The dispersion diagram is the plot of $\omega(\mathbf{k})$ against $\mathbf{k}$. It is usually given for a particular temperature, for it has been found that $\omega(\mathbf{k})$ can be temperature dependent. See Martel et al., 1967, and Fulton et al., 1994.] Attention has therefore tended to be focused on the question of how to account for the occurrence of a cooperative phase transition and how to use the dispersion results to determine something about the effective Hamiltonian.

§ 9.2 A Model Hamiltonian

It is clear from the diversity of spin-Hamiltonians for isolated magnetic ions that the study of large clusters of interacting spins is really the study of a large number of different problems. As a result the literature contains accounts of work on many model systems which, for the most part, have only one feature in common, that very few exact solutions, eigenvalues or eigenfunctions, are known (Keffer, 1966; Zeiger and Pratt, 1973). The simplest of all models is that of two spins of 1/2 coupled by a Heisenberg exchange interaction, which is a completely soluble problem. The Hamiltonian is

$$H = \mathbf{s}_1 \cdot \mathbf{s}_2 \tag{9.1}$$

and its eigenstates and eigenvalues are

$$|+,+\rangle, \frac{1}{\sqrt{2}}[|+,-\rangle + |-,+\rangle], |-,-\rangle : \frac{1}{4} \tag{9.2}$$

$$\frac{1}{\sqrt{2}}[|+,-\rangle - |-,+\rangle] : \frac{-3}{4}$$

where $+$ and $-$ denote eigenstates of m_s, and the first three states are eigenstates of $S = 1$ and the fourth is the eigenstate of $S = 0$. They can be classified as eigenstates of the total spin because $\mathbf{s}_1 \cdot \mathbf{s}_2$ commutes with all components of $\mathbf{S} = \mathbf{s}_1 + \mathbf{s}_2$. The model with three spins and Hamiltonian

$$[J_{12}\mathbf{s}_1 \cdot \mathbf{s}_2 + J_{23}\mathbf{s}_2 \cdot \mathbf{s}_3 + J_{31}\mathbf{s}_3 \cdot \mathbf{s}_1] \tag{9.3}$$

is also completely soluble. This can be predicted, for $D_{1/2} \times D_{1/2} \times D_{1/2} = 2D_{1/2} + D_{3/2}$. The states in $D_{3/2}$ can be written down by inspection, and two total spins of 1/2 imply that a quadratic will have to be solved. When all the exchange interactions have the same magnitude, J, the eigenvalues and eigenfunctions are

$$S = \tfrac{3}{2} : E_{3/2} = \tfrac{3}{4}J,$$
$$|+++\rangle,$$
$$\tfrac{1}{\sqrt{3}}[|-++\rangle + |+-+\rangle + |++-\rangle],$$
$$\tfrac{1}{\sqrt{3}}[|+--\rangle + |-+-\rangle + |--+\rangle],$$
$$[|---\rangle] \tag{9.4}$$
$$S = \tfrac{1}{2} : E_{1/2} = -\tfrac{1}{4}J,$$
$$\tfrac{1}{\sqrt{3}}[|-++\rangle + \omega|+-+\rangle + \omega^2|++-\rangle],$$
$$-\tfrac{1}{\sqrt{3}}[|+--\rangle + \omega|-+-\rangle + \omega^2|--+\rangle] :$$

$$\tfrac{1}{\sqrt{3}}[|-++\rangle + \omega^2|+-+\rangle + \omega|--+\rangle],$$
$$-\tfrac{1}{\sqrt{3}}[|+--\rangle + \omega^2|-+-\rangle + \omega|--+\rangle].$$

The two $S = 1/2$ levels coincide in energy because the Hamiltonian is invariant under a "rotation group," one element of which is the operation $1 \rightarrow 2, 2 \rightarrow 3$, $3 \rightarrow 1$. With ω equal to $\exp(2\pi i/3)$ the effect of the symmetry operation is easy to see. (The phases of the states with $M_S = -1/2$ are chosen to that S_- acting on a state with $M_S = 1/2$ gives the state with $M_S = -1/2$.) The eigenvalues can be found using the relation

$$
\begin{aligned}
\mathbf{s}_1 \cdot \mathbf{s}_2 + \mathbf{s}_2 \cdot \mathbf{s}_3 + \mathbf{s}_3 \cdot \mathbf{s}_1 &= \frac{1}{2}[\mathbf{S} \cdot \mathbf{S} - \mathbf{s}_1 \cdot \mathbf{s}_1 - \mathbf{s}_2 \cdot \mathbf{s}_2 - \mathbf{s}_3 \cdot \mathbf{s}_3] \\
&= \frac{1}{2}\left[S(S+1) - 3 \cdot \frac{1}{2} \cdot \frac{3}{2} \right] \quad (9.5)
\end{aligned}
$$

The sum of the eigenvalues is zero, as it would also have been if the exchange interactions had all been different. The result, for finite matrices, that $Tr AB = Tr BA$ is useful in proving such relations. On replacing s_z by $-i(s_x s_y - s_y s_x)$ it immediately follows that the trace of s_z and so, by symmetry, the trace of any component of a single spin, is zero. When the exchange constants have arbitrary values the eigenvalues are

$$
\begin{aligned}
S = 1, \quad & E_1 = \frac{1}{4}[J_{12} + J_{23} + J_{31}], \\
S = 0, \quad & E_\pm = -\frac{1}{4}[J_{12} + J_{23} + J_{31}] \\
& \pm \frac{1}{8}\sqrt{(J_{12} - J_{23})^2 + (J_{23} - J_{31})^2 + (J_{31} - J_{12})^2} \quad (9.6)
\end{aligned}
$$

An observation of the energy level spacings will give two quantities, $[J_{12}+J_{23}+J_{31}]$ and $[(J_{12}-J_{23})^2+(J_{23}-J_{31})^2+(J_{23}-J_{31})^2]$, which, perhaps surprisingly, are not sufficient to determine unique values for the three J's. With larger numbers of exchange-coupled spins the determination of the eigenvalues and eigenfunctions becomes increasing difficult, so usually the models are chosen to have symmetry properties that can be expected to help (e.g., linear chains of equally spaced ions or ions in larger and larger clusters having cubic symmetry).

The original motivation for the study of exchange-coupled spin systems arose from the Heisenberg-Dirac form for exchange interactions, which seemed to offer the possibility of understanding ferromagnetism and the change in physical properties that occurs on passing through the Curie temperature. The emphasis has now changed, partly because most of the ferromagnetic examples are metallic and there is some doubt about whether the H-D model applies, and partly because in most of the systems in which the presence of localized spins has been confirmed the isotropic part of the exchange interaction has the "wrong" sign and leads to antiferromagnetism. From the academic point of view, the arrangement of the eigenvalues does not depend on changing the sign of the exchange, for doing so simply inverts the whole energy level pattern. But when finding any eigenvalues, even approximately, is a major problem it is unrealistic to approach the two cases in this way, particularly as the thermodynamic properties depend primarily on the distribution of the low-lying energy levels.

Nevertheless, a good deal of effort continues to be put into finding the eigen-values and eigenstates of clusters of spins, where symmetry properties and the isotropic nature of the interaction can be exploited (see Caspers, 1989).

There is another problem with a finite number of spins, that of defining what is meant by a phase change. They are usually regarded, theoretically, as associated with singularities in thermodynamic functions. In the Heisenberg model, for example, every eigenvalue will be proportional to J, which will therefore appear in the thermodynamic functions as J/kT. It is often conve-nient to set $J/kT = z$ and regard z as a complex variable which reduces to a physically important parameter on the real axis (Rushbrooke et al., 1974). The theory of complex variables can then be used to manipulate expressions, partic-ularly power series expansions and Padé approximants, of the thermodynamic functions when written as functions of z. Phase changes are associated with singularities in these functions. The difficulty with a finite number of spins stems from the form of the partition function, which is a sum of exponentials, $\exp(-E_n/kT)$, where n ranges over the eigenvalues with their degeneracies. Each E_n/kT becomes a multiple of z and any expression of the form $\exp(-az)$, when expanded in a power series in z, is absolutely convergent. So is any finite sum of such expansions. Since a finite number of spins has a finite number of energy eigenvalues, the expansion of its partition function has no singularities and it follows that there are no phase transitions. If singularities are to emerge it is necessary to let the number of spins tend to infinity.

The experimental position is rather different, for there is no a priori reason to assume that the model of isotropically exchange-coupled spins is an accurate description of the system that is being studied, though there is often some expectation, usually based on the EPR properties of isolated ions, that some systems are closer to the model than others. Even so, with large assemblies of coupled spins there are no experimental techniques which will allow an assumed effective Hamiltonian to be thoroughly checked. Of the techniques that are available, inelastic neutron scattering at low temperatures is the one that has probably given most information (Fulde and Loewenhaupt, 1985). It is similar to the Raman process. A beam of monoenergetic neutrons, which is regarded as rather like a monochromatic wave, is found to induce transitions in the spin system, usually from a lower- to a higher-energy state, so that the wave loses some energy, but not all of it. The emergent beam therefore has a reduced energy and a different direction. The energy loss and the change in direction can be measured, showing that a wavelike mode of a definite energy and having a definite direction of propagation has been excited in the spin system. (There is conservation of energy and momentum.) By using a range of incident neutron energies and directions it is possible to build up what is known as a spin-wave spectrum, the $\mathbf{k}$, $\omega(\mathbf{k})$ dispersion relation, for the modes that have been excited in the sample. As with the spin-Hamiltonian for an isolated magnetic ion, it greatly simplifies expressing the experimental results if they can be shown to be consistent with the properties that emerge from an effective Hamiltonian, particularly if it can also be shown that the temperature of the phase transition emerges from the same effective Hamiltonian.

§ 9.3 Phase Transitions in General

One of the everyday examples of a phase transition is the transition from a gaseous to a liquid phase as the temperature is lowered, the most striking feature being the remarkable change in physical properties which occurs as a result of a very small temperature change in the vicinity of a particular temperature, the boiling point. If an explanation of this phenomenon is to emerge from a quantum mechanical model there is an immediate difficulty, that in choosing the generic Hamiltonian, at least along the lines that have been used in previous chapters, it would seem that it ought to be rotationally invariant. Yet it is well known that a gas in a container condenses into a liquid at the bottom of the container, which is hardly consistent with rotational symmetry, particularly if the gas is in a spherical container. The explanation of why the droplet is at the bottom is obvious, for it is due to gravity, the effect of which is not usually included in any Hamiltonian.

If the liquid is further cooled it will form a solid, which is likely to be an agglomeration of small crystals, all of which have the same crystal structure. For any given crystallite there is then a question of what has determined the directions of its crystallographic axes. Assuming the liquid is of infinite extent there seems no reason why specific directions should emerge.

Similar questions arise with ferromagnets. At high temperatures they are usually paramagnetic and no macroscopic magnetic moments exist unless an external magnetic field is applied. But, as the temperature is lowered, it is often stated that there comes a critical value below which the sample begins to show a permanent magnetic moment in a specific direction. A phase change from para- to ferromagnetism has taken place. In fact, this description of the phase change is inaccurate, as Weiss realized many years ago, for his freshly prepared iron had no moment below the phase transition. To develop the moment he found that it was necessary to take the sample round part of a hysteresis cycle, and the picture which then emerged was that the sample consisted initially of many crystallites, with random directions for their crystallographic axes, with each crystallite having a magnetic moment in a direction determined by its crystallographic axes. The production of a macroscopic moment only came when the moments in the individual crystallites were reoriented into equivalent crystallographic directions that were closest to the direction of the applied magnetic field. Nevertheless, in the as-prepared sample it still seemed that each crystallite was magnetized in a specific direction. So what was it that determined the direction? A tentative explanation could have been that it was decided by the directions of magnetization of the neighboring crystallites, but this is not really satisfactory because it only transfers the problem to that of deciding what had determined the directions of their magnetizations.

The problem of understanding how phase transitions arise is one of considerable difficulty, and to make progress it seems inevitable that assumptions have to be made, which raises the question of whether or not they are valid. Here it will be assumed that as a phase transition is approached the system becomes particularly sensitive to interactions that have been omitted from the generic Hamiltonian, interactions which have little effect on the temperature at which the phase transition occurs but which may have a large macroscopic effect on the structures that occur on either side of it, so that there are different effective Hamiltonians on the two sides. At least one, and perhaps both, of these should not be confused with the generic Hamiltonian, *which will be regarded as unchanged in moving from one phase to another and which will therefore be rotationally invariant.* The theoretical explanation of a phase transition might then be expected to emerge from a limiting process in a mathematical theory in which it may be necessary to have symmetry-breaking interactions which can be allowed to tend to zero after some other limit has been taken.

To support these assumptions a liquid to solid phase transition will be considered using two conceptual eigenstates of the low temperature phase. The first, $|\mathbf{R}, \alpha, \theta, \phi, n\rangle$, is regarded as an eigenstate of an effective Hamiltonian, H_c, which breaks the rotational symmetry and describes the solid by giving the position, $\mathbf{R}$, of its center of mass and its orientation by specifying three angular variables, α, β, and ϕ. (The direction of a line through the origin relative to axes in space is determined by two angles, α and β. A third, ϕ, is needed to describe a rotation about this line.) A second state,

$$|\mathbf{R} + \delta\mathbf{R}, \alpha + \delta\alpha, \beta + \delta\beta, \phi + \delta\phi, m\rangle,$$

describes a similar eigenstate of a related symmetry-breaking H_c, one that describes the crystal when it has been displaced and rotated by a small amount. Depending on the size of the crystal, the magnitudes of the matrix elements of any operator in the generic Hamiltonian, the underlying rotationally invariant Hamiltonian, when taken between $|\mathbf{R}, \alpha, \beta, \phi, n\rangle$ and $|\mathbf{R} + \delta\mathbf{R}, \alpha + \delta\alpha, \beta + \delta\beta, \phi + \delta\phi, m\rangle$ can be expected to decrease rapidly with increases in any of $\delta\mathbf{R}$, $\delta\alpha$, $\delta\beta$, and $\delta\phi$, for many reasons, one being the decrease in the overlaps of the wave functions that describe the vibrational motions of the nuclei. Indeed, if the relative displacements are already such that the matrix elements of operators in the generic Hamiltonian, taken between the eigenstates of the two H_c's, for all their thermally populated states, can be regarded as zero, the fact that both sets of states belong to the same Hilbert space, that of the generic Hamiltonian, leads to the conclusion that the eigenstates of both H_c's can be regarded as eigenstates of the generic Hamiltonian. The reasoning can be extended by introducing a whole family of H_c's, each of which will have the same pattern of eigenvalues, with similar sets of orthogonal states. The operators that might, but do not, connect the families, can then be extended to include finite multiples of the operators in the generic Hamiltonian. (The idea here is that as the size of the system increases any finite product of annihilation and creation operators is increasingly unlikely to couple the states of different H_c's.) It will then be much more convenient to use just a single H_c, which describes a crystal at a particular orientation, determined by some outside influence, than to use the generic Hamiltonian. Both will have the same low-lying eigenvalues, with those of the generic Hamiltonian having a much higher degeneracy, due to the multiplicity of equivalent H_c's. The above supposition allows the introduction of the concept of *strongly orthogonal states*, many-particle states which are so different from one another that the usual operators, those that produce rearrangements of finite numbers of particles, in this case electrons and nuclei, will have zero matrix elements between any of their bras and kets. (There are operations, such as the bulk rotation of a cubic crystal placed on a horizontal plane, which have to be excluded, for the operator that describes such a rotation clearly does have matrix elements between strongly orthogonal states.) Below the phase transition it is as if the nuclei are providing a framework for each H_c. Above the phase transition the nuclei no longer provide such a framework, so the rotationally invariant generic Hamiltonian should then be used.

The above reasoning has some similarity with the "broken symmetry" concept (Palmer, 1982). This is often introduced using the Heisenberg model of a ferromagnet, N spins of $1/2$, for which the ground state has $S = N/2$ with degeneracy $N + 1$ (see the next section). Each state in this family is an eigenstate of the model Hamiltonian and none of them have rotational symmetry. The reason is that the family span a $D_{N/2}$ i.r. of the rotational group in spin space and any state that is rotationally invariant necessarily transforms as D_0. If an external magnetic field is applied, the degeneracy is lifted and the lowest state remains an eigenstate, with its magnetic moment lined up in the direction of the field. When the field is then reduced to zero the state remains unaltered, for it is an eigenstate throughout the field cycle, and the result is a ferromagnetic in zero external field with an aligned moment. The symmetry of the Hamiltonian has not been broken, but to an observer the symmetry would appear to have been. The same sort of reasoning can be applied to any set of degenerate states that span an i.r. There is a problem, though, as can be illustrated with the ferromagnet, with the assumption that the moment is coupled to a cubic lattice structure. It then seems reasonable to assume that, following the same field sequence, the moment will be found to be aligned along a crystallographic direction, for which there are six possibilities. It is now of interest to apply the same reasoning to an excited energy level. If the states in this span a different i.r. of the cubic group, it is not obvious that the effect of the magnetic field cycle, even if it does leave the system in a "broken symmetry" state, will be a state which is "broken" in the same way. Maybe the moment will be found to be aligned along a threefold rather than a fourfold axis. It would therefore seem that for a system that has many energy levels that can be thermally occupied at temperatures below that of the phase transition it is not enough to base the "symmetry-breaking" argument on a property of just one degenerate family.

Turning back to Chapter 7 it can now be seen that a restriction can be removed. In §7.1, it was assumed that in the generic Hamiltonian the nuclei are at fixed positions and then, in §7.5, that the nuclei can vibrate about lattice sites. In the light of the above considerations it is now possible to take the further step to a generic Hamiltonian in which the nuclear motion is unrestrained, by replacing the previous assumptions by the assumption that at some temperature above that of interest there has been a phase change and that below it a symmetry-broken Hamiltonian can be used in place of the full Hamiltonian.

§ 9.4 Ferromagnetism

Although there is a shortage of examples of spin systems that show ferromagnetism, the model of spins of $1/2$ coupled by ferromagnetic isotropic exchange interactions between nearest neighbors has been widely studied. (There are also theories for the conducting ferromagnets, such as iron, which are directed towards producing effective Hamiltonians of this form, but in reciprocal space rather than actual space, see Moriya, 1985.) The Hamiltonian is

$$- |J| \sum_{\langle i,j \rangle} \mathbf{s}_i \cdot \mathbf{s}_j, \qquad (9.7)$$

where $\langle i, j \rangle$ indicates that the summation is over adjacent pairs of ions with $s = 1/2$. It is immediately obvious, writing $\mathbf{s}_i \cdot \mathbf{s}_j$ as $(1/2)[s_{i+}s_{j-} + s_{i-}s_{j+}] + s_{iz}s_{jz}$, that the state $|+, +, +, \ldots\rangle$, in which all the spins have $m_s = 1/2$, is an eigenstate with the eigenvalue $-1/4Nz|J|$, where N is the number of spins and z is half the number of nearest neighbors of each spin (the number of pairs is Nz, forgetting that boundary spins have a different number of neighbors). Further, operating on this state with S_+, the sum of the individual s_+ operators, gives zero, showing that the state belongs to $S = N/2$ and has $M_S = N/2$. Since each operator in (9.7) commutes with the total spin, it follows that the whole operator also commutes with total spin. There will therefore be $(2S+1)$ degenerate states with this energy, differing in their M_S values, which range in integer steps from $-N/2$ to $N/2$. They form a basis for a D_S irreducible representation of the rotation group in the spin space. A simple way to find them is to use the S_- operator repeatedly. For example, the state with $M_S = (1/2)(N-2)$, which is obtained by applying S_- to $|+++++\rangle$ followed by normalization, is

$$\frac{1}{\sqrt{N}}[|-+++++\ldots\rangle+|+-++++\ldots\rangle+|++-+++\ldots\rangle$$
$$+|+++-++\ldots\rangle+\ldots]. \qquad (9.8)$$

A further application of S_-, gives the state with $M_S = (1/2)(N-4)$, and so on. However, the task of determining the states with steadily decreasing values for M_S becomes increasingly tedious and is seldom necessary, for all that is needed can usually be deduced by replacing a state with $M_S = (1/2)(N-2r)$ by $(S_-)^r |+++++\rangle$ multiplied by a normalizing factor.

If periodic boundary conditions are imposed a few further eigenstates and eigenfunctions can be found. These are the states

$$(S_-)_\mathbf{k}|+,+,+,\ldots\rangle = \frac{1}{\sqrt{N}}\left[\sum_j \exp(i\mathbf{k}\cdot\mathbf{r}_j)(s_{-j})\right]|+,+,+,\ldots\rangle, \qquad (9.9)$$

where $\mathbf{k}$ is a vector that ranges over the reduced Brillouin zone and $\mathbf{r}_j$ is the position of the jth spin. They form an orthonormal set, each state of which has $M_S = (1/2)(N-2)$. The state obtained with $\mathbf{k} = 0$ is identical to the state given in (9.8), so it belongs to $S = N/2$. All the others are eigenstates of total spin $S = (1/2)(N-2)$. (This can be shown by applying S_+. For any choice $\mathbf{k} \neq 0$ it gives zero.) There is thus a band of $(N-1)$ states each having $\mathbf{k} \neq 0$ and the same S. Their eigenvalues are

$$E_k = -|J|\left[\frac{z}{4}(N-4) + \frac{1}{2}\sum_t \exp(i\mathbf{k}\cdot\mathbf{t})\right], \qquad (9.10)$$

where $\mathbf{t}$ ranges over the vectors that join one lattice site to its neighbors. The only other known eigenvalues seem to be those of special models, such as linear chains (see Caspers, 1989). The only other general result is that the highest energy level, which becomes the ground level when the sign of the exchange is reversed, has $S = 0$. It will be considered in the context of antiferromagnetism. It is unfortunate that successive applications of the $(S_-)_\mathbf{k}$ operators to $|+,+,+,\ldots\rangle$ does not give a sequence of eigenstates of (9.7) and that attempts to modify their definition so that the new versions do have this property have not, so far, been successful.

In the eigenstate $|+,+,+,\ldots\rangle$ the spin moments of the individual ions are aligned in the same direction, so if it had not been a state in a highly degenerate set it would have been tempting to regard it as describing the ferromagnetic ground state. This would have overlooked an unacceptable feature, that the direction of alignment has been determined by the choice of direction for the axis of quantization. Indeed, the degeneracy can be regarded as arising precisely because this unphysical possibility has not been avoided. It can be circumvented if it is assumed that the spin system is exposed to a small external magnetic field, such as that due to the earth, which will create a Zeeman splitting of all the degeneracies associated with nonzero total spins. There will then be an isolated $|+, +, +, \ldots\rangle$ ground state provided that the axis of quantization is taken either in the direction of the field or opposite to it, depending on the sign of the g factor which relates spin to magnetic moment. At high enough temperatures, when many excited states are thermally populated, the presence of such a field should be unimportant, in which case it can be concluded that the system should no longer be regarded as ferromagnetic. The interesting question is then whether or not the model predicts that there will be a phase change from a paramagnetic to a ferromagnetic phase as the temperature is lowered and the number of spins goes to infinity.

§ 9.5 Antiferromagnetism

The study of the Heisenberg Hamiltonian with the opposite sign of the exchange interaction has produced an extensive literature on approximation methods, for even less is known about its low-lying eigenvalues and states. Compared with the H-D model of ferromagnetism there is even more doubt about whether it describes any actual system, for at least with the ferromagnet there is some reason to suppose that the actual ground state will not differ substantially from the $|+ + + + + \ldots\rangle$ state, when the axis of quantization is correctly chosen. In the antiferromagnets one of the few exact results is that, for a lattice that can be divided into two sublattices, so that the spins on one sublattice are only coupled to spins on the other sublattice, the ground state has a total spin of zero (Lieb and Mattis, 1962). This result is particularly difficult to reconcile with experimental observations.

At first sight, the usual experimental picture, that the ground state can be written in the form $|+ - + - + - \ldots\rangle$, (with the plus signs for spins on the A sublattice and the minus signs for spins on the B sublattice, using an axis of quantization determined experimentally) has total $M_s = 0$, which is compatible with $S = 0$. The first difficulty is, however, that the above state is not an eigenstate of total spin, so there is no possibility of its being an eigenstate of the Heisenberg Hamiltonian. The second is that if it had been a state with $S = 0$ the magnetic moment at every site would have been zero. (This follows from the Wigner-Eckart theorem, that the matrix element of any component of a vector taken between states both of which have $S = 0$ is zero, for $D_0 \times D_0$ does not contain D_1.) What is even more disturbing is that although the state $|+ - + - + - \ldots\rangle$ does have $M_S = 0$ it is very far from being an eigenstate of $S = 0$. This may be seen by writing the above ground state in a form that has all the A spins written first with the B spins following, so that it appears as

$|++++...----...\rangle$. A state of the A spins alone, in which all the spins have the $+$ sign, is a state which has $M_S = N/2$, where N is the total number of A spins. It is also an eigenstate of total spin, with $S = N/2$. Similarly, the state of the B spins, all of which are in the $-$ state, is a state of another $S = N/2$ with $M = -N/2$. At this point the notation is becoming clumsy, so to avoid confusion T will be used instead of S to describe the states of the B spins. The ground state of the experimentalist can then be expressed as $S = N/2$, $M_S = N/2$, $T = N/2$, $M_T = -N/2$. Now when a manifold of spin $S = N/2$ is coupled to another manifold with spin $T = N/2$ the totality of states can be written as eigenstates of a total spin, which will be denoted by J, which takes the values N, $(N-1)$, $(N-2)$, ..., 0 (from $D_S \times D_T = D_{S+T} + D_{S+T-1} + ... D_0$). Noting that D_0 occurs only once, it follows that there is only *one* state in this assembly of product states that has a total spin of zero. So the experimental ground state will only be found in one state which has $J = 0$ for its total spin, and this state is easily constructed using angular momentum operators. It is

$$|J = 0, M_J = 0\rangle = \frac{1}{\sqrt{N}}\left[\left|\frac{N}{2}, -\frac{N}{2}\right\rangle - \left|\frac{N}{2} - 1, -\frac{N}{2} + 1\right\rangle \right.$$
$$\left. + \left|\frac{N}{2} - 2, -\frac{N}{2} + 2\right\rangle - ...\right], \qquad (9.11)$$

where in each ket on the right-hand side the first symbol, of the form $(1/2)N - r$, where r ranges from 0 to $N/2$, denotes that the state has a total S equal to $(1/2)N - r$ with a projection onto O_z that is equal to its total spin. That is, the moment is fully aligned. The second symbol refers to the T spins and describes a state with a total spin of $(1/2)N - r$ with a projection of $-(1/2)N + r$ onto O_z. The T moment is therefore aligned antiparallel to the S moment. Thus the total aligned moment for each component state is zero, and the amplitude of the particular state $|++++...----...\rangle$ $(r = 0)$ in the $J = 0$ state is $1/\sqrt{N}$. Since there are many other orthogonal states in the totality of states that have $J = 0$ none of which contain $|++++...----...\rangle$ it follows that, since the $J = 0$ state that does contain $|++++...----...\rangle$ does so in a negligible way as N becomes large, the ground state of the Heisenberg Hamiltonian, whatever combination of $S = 0$ states it contains, contains a negligible amount of the $|++++...----...\rangle$ state.

There are several ways in which it may be possible to circumvent the above difficulty, one being that there is a misinterpretation of the experimental results and that the observed local moments do not correspond to having each spin fully aligned. The expression "spin fluctuation" occurs fairly frequently in the literature. It is often used to express the idea that the spin at each site is fluctuating in time so that its temporal value averages to less than 1/2. Eigenstates are, however, stationary states, so there is no associated time dependence such as is implied by the above description. A better description of the basic idea would seem to be that there are extra terms in the effective Hamiltonian which connect a $(+, -)$ arrangement of spins on adjacent sites to a $(-, +)$ arrangement, which will result in there being an apparent reduction in the expectation values of s_z for spins on the A sublattice and an increase for the spins on the B sublattice. (There are already operators in the H-D Hamiltonian of the required nature, so this suggestion is tantamount to proposing an increase in their effect by introducing anisotropic exchange interactions—which are almost certainly present.)

There is another argument against the H-D model, which is the existence of ferrimagnets (Standley, 1962; Tebble and Craik, 1969). Many of them have complicated magnetic structures, though the basic concept is the same as that used for antiferromagnets, that of two sublattices with antiparallel spin alignments. The difference is that the ions on the A sublattice have different magnetic moments from those on the B sublattice, so the antiparallel alignments give a net magnetization. (They are of considerable technical importance because they are insulators with macroscopic magnetic moments that are comparable with those of the ferromagnetic metals.) This would not occur if the low-lying states were eigenstates of low total spins, with zero for the ground state. (The H-D model ignores the possibility that effective spins of $1/2$ may have different magnetic moments.) There is therefore a good deal of evidence that points to the need to use more realistic descriptions of the interactions between spins. Unfortunately, it is difficult enough to make progress with the Heisenberg model, so the prospect of having to use a Hamiltonian with much less symmetry is not attractive.

There is also some reason to suppose that the H-D model is, in a sense, unstable. Suppose a small external magnetic field is applied along the Oz direction. A term of the form $g\beta B J_z$ should then be added to the Hamiltonian. This interaction commutes with the Hamiltonian, so it will produce a Zeeman splitting of every manifold of total spin except those that have $J = 0$, which includes the ground level. If therefore there is any nearby manifold that has J greater than zero a state split off from it may cross the ground $J = 0$ state. Indeed, as the number of spins tends to infinity it is probable that there will be many manifolds with a variety of total spins greater than zero which are close enough for many of the Zeeman-split states to cross the ground $J = 0$ state. There seems to be no experimental evidence that antiferromagnets are unstable under the application of small magnetic fields, which suggests that not only should the interactions be anisotropic but that these will counteract this type of instability.

§ 9.6 An Application of the Lanczos Method

There is a potentially useful method for studying antiferromagnetism, based on the Lanczos method. An initial normalized state, $|1\rangle$, is chosen. The Hamiltonian operator is then applied to it. This will produce a multiple of $|1\rangle$ and a remainder, which is orthogonal to $|1\rangle$. When normalized the remainder can be denoted by $|2\rangle$. The Hamiltonian is then applied to $|2\rangle$, to produce a linear combination of $|1\rangle$ and $|2\rangle$ together with a remainder. The remainder is again normalized to give a state, $|3\rangle$, which is orthogonal to both $|1\rangle$ and $|2\rangle$. The process, as it is continued, produces a sequence of orthonormal states with which to set up the matrix of the Hamiltonian. If this can then be diagonalized it will give eigenstates some of which will contain $|1\rangle$. Indeed, if the initial choice is at all like the true ground state the method should, in principle, generate the actual ground state in what appears to be an economical way, by avoiding the introduction of unnecessary states.

The method was examined in Stevens (1992) using the Heisenberg Hamiltonian and beginning with the state in (9.11). As this already had a total spin of zero the technique could only produce further states with zero spin. Even so the method soon became almost unmanageable, so it was abandoned when it was realized that the Heisenberg model was probably not the model that should be studied. Nevertheless, it did suggest that an inspired choice of $|1\rangle$ might lead to progress. It would therefore seem of interest to begin with a more realistic Hamiltonian and choose for $|1\rangle$ the state in which the A spins have $m_s = 1/2$ and the B spins have $m_t = -1/2$, paying particular attention to the operators of type $(s_-)_i (t_+)_j$ which flip spins on adjacent sites.

On applying such a Hamiltonian to $|1\rangle$ these operators introduce states in which the moment on the A sublattice has been decreased by unity and that of the B sublattice has been increased by unity, states that are not present in $|1\rangle$. It follows that they must therefore be in $|2\rangle$. On applying the Hamiltonian to $|2\rangle$ the same operators produce more spin reversals, so generating states which are not in either $|1\rangle$ or $|2\rangle$. These states must therefore belong to $|3\rangle$. Continuing the process, it follows that each member of the set of orthonormal states $|1\rangle$, $|2\rangle$, $|3\rangle$, ... can be characterized by the lowest value of M_S that it contains. Each state $|n\rangle$ is therefore a sum of product states with a range of M_S and M_T values, where $M_T = -M_S$. The value of n can then be related to the smallest occurring M_S value by the relation that the smallest M_S equals $(1/2)(N + 1 - n)$. There seems no reason why the sequence which generates $|1\rangle$, $|2\rangle$, ... should end before a final state, one in which M_S takes the lowest possible value, which is $-(1/2)N$, has been reached. A state in which the highest value of M_T, $(1/2)N$, has then also been reached. This state, which is reached after N steps, can be denoted $|n = N + 1\rangle$. It has a particularly simple structure for it is the state obtained from $|1\rangle$ by reversing all the spins. In fact, whichever state $|n\rangle$ is examined, it can be seen that there is another state belonging to the set that can be obtained from it by reversing all the spins. This related state can be denoted $|n'\rangle$, where $n' = N + 2 - n$. On diagonalizing the matrix of the Hamiltonian within this family of states, it follows that in each eigenstate the state labeled $|n\rangle$ will be accompanied by the state $|n'\rangle$. As the Hamiltonian is time-reversal invariant, each of its eigenstates will either be odd or even under time reversal, and to achieve this each $|n\rangle$ must be accompanied by an $|n'\rangle$ with the same modulus but possibly a different phase. It follows that each eigenstate can be written in the form

$$\frac{1}{\sqrt{2}}[a_1|1\rangle + b_1|1'\rangle] + \frac{1}{\sqrt{2}}[a_2[|2\rangle + b_2|2'\rangle] + \frac{1}{\sqrt{2}}[a_3|3\rangle + b_3|3'\rangle] + \ldots \quad (9.12)$$

where $|a_n| = |b_n|$. Such a state has quite a different structure from the state usually used to describe the experimentally observed ground state, $|+, -, +, -, \ldots\rangle$.

Several problems now arise. One is the question of how it is possible, on cooling a magnetic crystal, for it to have a transition into a state that is neither odd nor even under time-reversal symmetry, a question which can be rephrased into: What is it that causes the spin at a particular site to orient in a particular direction? In the case of a ferro- or ferrimagnet there is no basic problem, for there will be the earth's field, if needed. But for an antiferromagnet it seems that something which distinguishes between two adjacent sites is needed and it is not obvious what it can be. So the question arises as to whether the usual description is correct.

There are several points that may be relevant. One is that it may be incorrect to assume that in the ground state of an antiferromagnet a particular site has its moment in a particular direction, for, on closer examination, this does not seem to have been established by experiment. What the experiments have established are the magnitudes of the magnetic moments at sites and the correlations in directions of moments at different sites. If one site has a moment in an up direction the correlation with an adjacent site may well show that its moment is in the opposite direction, which is not the same as establishing a definite direction for both sites.

Another point can be illustrated by making a slight change in the state given in (9.12), by changing the phase of each $|n'\rangle$ state, if necessary, so that each pair in it is a multiple of a combination of the form $|n\rangle + |n'\rangle$. The pair corresponding to $n = 1$ contains the state $|1\rangle$ in which all the spins on the A sublattice have $m_s = 1/2$ and all the spins on the B sublattice have $m_t = -1/2$ and a second state $|1'\rangle$ in which all the spins are reversed. $|1\rangle$ and $|1'\rangle$ are therefore strongly orthogonal. Similarly, $|2\rangle + |2'\rangle$ is a combination of two states that are also strongly orthogonal and, furthermore, $|1'\rangle$ is strongly orthogonal to $|2\rangle$ and $|1\rangle$ to $|2'\rangle$. These strongly orthogonal properties continue as n increases and it is only when n is of the order of $N/2$ that the two states in an $|n\rangle + |n'\rangle$ combination cease to be strongly orthogonal to one another and to adjacent

$|n\rangle + |n'\rangle$ combinations. Also, from the nature of the experimental ground state it seems that it is dominated by states that do not depart far, in the numbers of reversed spins, from the $|+, -, +, -, +, \ldots\rangle$ state, which suggests that the coefficients of the a_n coefficients in (9.12) are large for low values of n and fall to very small values as n approaches $N/2$. It then appears that any state which does not violate time-reversal symmetry can be regarded as composed of pairs that are not only strongly orthogonal within the pairs but are strongly orthogonal to most other pairs. The only exceptions will be the pairs which have n near $N/2$, pairs which have very small a_n coefficients, so they are of negligible importance. It follows that (9.12) can be broken into two parts:

$$a_1|1\rangle + a_2|2\rangle + a_3|3\rangle + \ldots,$$

and
$$(9.13)$$

$$a_1|1'\rangle + a_2|2'\rangle + a_3|3'\rangle + \ldots,$$

both of which break the time-reversal symmetry requirement, with each having the property that as far as most calculations are concerned either one is as good as the other, and just as good as the states obtained by combining them together into states that do have the required symmetry. This result is similar to, but not quite identical with, the "strong-orthogonality" argument used previously.

There it has been assumed that the gas-liquid, liquid-crystal, and para-ferromagnetic phase changes occur without change in the basic Hamiltonian, and that in each case it is possible to use an effective Hamiltonian which has a lower symmetry, is more convenient to use, and is just as viable. These effective Hamiltonians do not break time-reversal symmetry. The antiferromagnet is different, because the ground state determined by experiment does appear to break time-reversal symmetry. It has therefore seemed necessary to seek a way of seeing how an effective Hamiltonian that does break time reversal symmetry can arise as a substitute for one that does not.

§ 9.7 Spin waves in Antiferromagnets

A common assumption in the treatment of cooperative systems is that the low-lying states can be obtained as "excitations" from a ground state. A good deal of effort has therefore been devoted to finding operators which, acting on the ground state, create excitations. Indeed, this activity often appears to be of more interest than determining the nature of the ground state, particularly if the operators can be assumed to have familiar commutation rules. In a ferromagnet the excitations are pictured as spin reversals that propagate through the lattice. For the two-sublattice structure of an antiferromagnet a more complicated but related picture has been developed. Many papers seem to begin with the H-D antiferromagnetic exchange model and then replace it by a model in which each spin interacts with an internal magnetic field, which produces a ground state with the alternating spin arrangement. In the light of some of the comments already made, it would seem better either to begin with the former and include anisotropic exchange interactions and zero-field splittings, if the spins are greater than $1/2$, or to go straight to the latter with its alternating internal magnetic field. The analysis commonly followed can cope just as well with either.

The general procedure is most conveniently illustrated by choosing spins of $1/2$. The magnetic structure is divided into two and the ground state is assumed to be the state in which all the spins on the A sublattice have $m_s = 1/2$ and all those on the B sublattice have $m_t = -1/2$. The exchange interaction is then regarded as coupling A and B spins on adjacent sites. A substitution introduced by Holstein and Primakoff (1940), or a modified version of it, is then invoked.

The basic idea is to replace the spin operators at a given site by operators that have slightly different and more convenient commutation rules, the creation and annihilation operators used in the study of harmonic oscillators. For a single spin of magnitude s the replacement is

$$
\begin{aligned}
s_z &= s - a^*a, \\
s_+ &= \sqrt{2s}\left(1 - \frac{a^*a}{2s}\right)^{1/2} a, \\
s_- &= \sqrt{2s}a^*\left(1 - \frac{a^*a}{2s}\right)^{1/2},
\end{aligned}
\tag{9.14}
$$

where a and a^* satisfy the commutation rule $aa^* - a^*a = 1$. (The square root of an operator is defined by its series expansion. In normal algebra such expansions are only convergent under certain conditions. For operators it is less obvious what these are. For what it is worth, it can be claimed that such an expansion is valid when $a^*a/2s$ is, in some sense, small. There is no doubt, though, that some of the expectation values of a^*a are far from small. So the hope must be that those which enter are not of this type.) These replacements preserve the commutation rules for the spin components for all values of the total spin, but not the magnitude of the total spin. They also vastly extend the dimensions of the Hilbert space. It is also relevant that the expression for s_z has quite a different form from that for s_x and s_y, which presumably incorporates the feature that the z direction is special. As it contains a^*a, which ought to remain small, there is an implication that the spin moment is close to being aligned in the z direction. For a ferromagnet this direction should be that of the macroscopic moment. For the antiferromagnet the z direction will be taken in the same direction for both A and B spins and the physically reversed direction for the B spins will be taken into account by using b and b^* operators in the Holstein-Primakoff substitution, with the substitution written in a modified form:

$$t_z = -t + b^*b,$$

$$t_- = \sqrt{2t}b^*\left(1 - \frac{b^*b}{2t}\right)^{1/2} \qquad (9.15)$$

$$t_+ = \sqrt{2t}\left(1 - \frac{b^*b}{2t}\right)^{1/2}b,$$

with t replacing s. For spins of $1/2$ and a general form for the exchange interaction, the algebra becomes quite lengthy and it is even more so for larger spins with zero-field splittings.

For an adjacent pair that lie on the z axis the interaction will be taken to be $J_1 s_z t_z + J_2[s_x t_x + s_y t_y]$. With the H-P replacement it becomes

$$J_1(1/2 - a^*a)(-1/2 + b^*b)$$
$$+ \quad (1/2)J_2[(1 - a^*a)^{1/2}ab^*(1 - b^*b)^{1/2}$$
$$+ \quad a^*(1 - a^*a)^{1/2}(1 - b^*b)^{1/2}b]. \qquad (9.16)$$

For a pair lying along the x direction the equivalent interaction is $J_1 s_x t_x + J_2[s_y t_y + s_z t_z]$, with a related form, obtained by substituting x by y, etc., for a pair lying along the y direction. These too are replaced by the H-P expressions. (It may be noted that because of the symmetry of the interaction there are no terms linear in the operators.) The spin-Hamiltonian is the sum of these interactions taken over all nearest-neighbor pairs. The next step is to expand the square roots and neglect all operators that contain products of a^*a and b^*b pairs. This removes a good many operators that would otherwise be awkward to deal with. The part that remains then has the form of a summation over products of just two operators, which can both be unstarred, both starred, or one starred and the other unstarred. This may be called a canonical form, for there is a standard way of dealing with it when the lattice is periodic.

The periodicity is used to separate the Hamiltonian into commuting parts, the first step being to introduce wavelike operators, by setting

$$a_{\mathbf{k}} = \frac{1}{\sqrt{N}} \sum_i \exp(i\mathbf{k} \cdot \mathbf{r}_i)a_i,$$

$$a_i = \frac{1}{\sqrt{N}} \sum_{\mathbf{k}} \exp(-i\mathbf{k} \cdot \mathbf{r}_i)a_{\mathbf{k}}, \qquad (9.17)$$

$$b_{\mathbf{k}} = \frac{1}{\sqrt{N}} \sum_j \exp(i\mathbf{k} \cdot \mathbf{r}_j)b_j,$$

$$b_j = \frac{1}{\sqrt{N}} \sum_{\mathbf{k}} \exp(-i\mathbf{k} \cdot \mathbf{r}_j)b_{\mathbf{k}},$$

where in the definitions **k** can be regarded as running over all the vectors in the Brillouin zone (BZ) of the A sublattice or all those in the BZ of the B sublattice, for the two are identical. $\mathbf{r}_i$ and $\mathbf{r}_j$ range, respectively, over the A and B sites. The $a_{\mathbf{k}}$, $a_{\mathbf{k}}^*$ etc., operators obey the standard harmonic oscillator commutation rules, $aa^* - a^*a = 1$ for identical **k** values and $aa^* - a^*a = 0$ for differing values, with all the operators of one sublattice commuting with those of the other. The definitions are chosen so that if a displacement is made that leaves each sublattice invariant, $a_{\mathbf{k}}$, $a_{-\mathbf{k}}^*$, $b_{\mathbf{k}}$, and $b_{-\mathbf{k}}$ have the same phase change, whereas $a_{\mathbf{k}}^*$, $a_{-\mathbf{k}}$, $b_{\mathbf{k}}^*$, and $b_{-\mathbf{k}}$ have the opposite one. When the site creation and annihilation operators are replaced by the wavelike operators the periodic boundary condition reduces double summations over **k**'s to summations over single **k**'s, so leading to summations of commuting expressions of the form

$$
\begin{aligned}
&A(\mathbf{k})[a_{\mathbf{k}}^* a_{\mathbf{k}} + b_{\mathbf{k}}^* b_{\mathbf{k}} + a_{-\mathbf{k}}^* a_{-\mathbf{k}} + b_{-\mathbf{k}}^* b_{-\mathbf{k}}] \\
&+ B(\mathbf{k})[a_{\mathbf{k}} b_{-\mathbf{k}} + a_{\mathbf{k}}^* b_{-k}^* + a_{-\mathbf{k}} b_{\mathbf{k}} + a_{-\mathbf{k}}^* b_{\mathbf{k}}^*] \qquad (9.18) \\
&+ C(\mathbf{k})[a_{\mathbf{k}}^* b_{\mathbf{k}} + a_{\mathbf{k}} b_{\mathbf{k}}^* + a_{-\mathbf{k}}^* b_{-\mathbf{k}} + a_{-\mathbf{k}} b_{-\mathbf{k}}^*],
\end{aligned}
$$

where the summations are now to be taken over half the **k** values in each Brillouin zone because each expression has been arranged to be unchanged if $-\mathbf{k}$ is substituted for **k**. The $A(\mathbf{k})$, $B(\mathbf{k})$, etc., coefficients are complicated expressions in J_1, J_2, and the components of **k**. In this example they are all real. It is not necessary, however, to give them explicitly, for the prime need is to show that with the appropriate manipulations (9.18) can be rewritten in a form in which **k** and $-\mathbf{k}$ are no longer mixed. One procedure relies on choosing the right linear combinations of the various starred and unstarred operators so that (9.18) takes a new form, a summation of four expressions two of which have the form $\hbar\omega_{\mathbf{k}}\alpha_{\mathbf{k}}^*\alpha_{\mathbf{k}}$ and the other two have $-\mathbf{k}$ in place of **k**. A simpler procedure is to begin by solving the equations of motion of the operators:

$$
\begin{aligned}
[a_{\mathbf{k}}, H] \;=\; & i\hbar\frac{\partial a_{\mathbf{k}}}{\partial t} = A a_{\mathbf{k}} + B b_{-\mathbf{k}}^* + C b_{\mathbf{k}}, \\
& i\hbar\frac{\partial b_{-\mathbf{k}}^*}{\partial t} = -A b_{-\mathbf{k}}^* - B a_{\mathbf{k}} - C a_{-\mathbf{k}}, \qquad (9.19) \\
& i\hbar\frac{\partial b_{\mathbf{k}}}{\partial t} = A b_{\mathbf{k}} + B a_{-\mathbf{k}}^* + C a_{\mathbf{k}}, \\
& i\hbar\frac{\partial a_{-\mathbf{k}}^*}{\partial t} = -A a_{-\mathbf{k}}^* + B b_{\mathbf{k}} - C b_{-\mathbf{k}}^*,
\end{aligned}
$$

for, on giving each the same $\exp(i\omega t)$ time dependence, a set of equations is produced that is easy to solve. They give

$$
(\hbar\omega)^2 = (A+C)^2 - B^2 \quad \text{and} \quad (\hbar\omega)^2 = (A-C)^2 - B^2.
$$

On the way it becomes fairly obvious that the introduction of combinations of the form

$$
\begin{aligned}
\alpha_{\mathbf{k}} &= \cos\theta[a_{\mathbf{k}} + b_{\mathbf{k}}] + \sin\theta[a^*_{-\mathbf{k}} + b^*_{-\mathbf{k}}], \\
\beta_{\mathbf{k}} &= \sin\theta[a_{\mathbf{k}} + b_{\mathbf{k}}] - \cos\theta[a^*_{-\mathbf{k}} + b^*_{-\mathbf{k}}], \\
\gamma_{\mathbf{k}} &= \cos\phi[a_{\mathbf{k}} - b_{\mathbf{k}}] + \sin\phi[a_{-\mathbf{k}} - b_{-\mathbf{k}}], \\
\delta_{\mathbf{k}} &= \sin\phi[a_{\mathbf{k}} - b_{\mathbf{k}}] - \cos\phi[a_{-\mathbf{k}} - b_{-\mathbf{k}}],
\end{aligned}
\tag{9.20}
$$

together with their complex conjugates, will lead, with appropriate choices for θ and ϕ, to the required separation. As the standard convention is that frequencies are always positive, the picture which emerges is that there are four wave motions with the same $|\mathbf{k}|$ but only two different frequencies. Two of the waves will travel in the $\mathbf{k}$ direction with different frequencies and the other two will be their reflections and travel in the $-\mathbf{k}$ direction. Each motion is described as that of a spin-wave mode, where "mode" is used to describe a harmonic oscillator. Each mode therefore has a ladder of equally spaced energy levels, the spacing being $\hbar\omega$, where ω is the angular frequency of the wave.

It is a common expectation that it should be possible to decompose the low-amplitude excitations of any macroscopic system into a superposition of wave motions, though whether or not this is actually possible in any specific example needs to be demonstrated. In the case of the antiferromagnets, in-elastic neutron scattering has provided a certain amount of support, though how far this goes towards showing that each mode has a ladder of equally spaced energy levels is less clear. Certainly there is not an infinite ladder for each mode, for the number of states in any spin-Hamiltonian is far too limited for this to be possible. It seems much more likely that the harmonic oscillator picture holds for only a limited number of excitations of each mode, which raises the question of whether or not its description as a harmonic oscillator is adequate at temperatures of the order of the para- to antiferromagnetic phase transition.

An interesting feature of spin-wave theory is that it has resulted in a description in which the excitations are of Bose-Einstein type, excitations that have arisen from a spin-Hamiltonian in which the spins at different sites satisfy Boltzmann statistics, a description which in turn originated with electrons that satisfy Fermi-Dirac statistics!

Before leaving the topic of spin-wave theory a few further comments can be made. In many neutron scattering experiments an external magnetic field is applied, which requires an extension of the H-P treatment. It then becomes apparent that the model has incorporated a number of convenient assumptions, which have to be modified. The first is that the direction of the spin alignment is known. An applied magnetic field is not necessarily in the alignment direction, which raises the question of whether the alignment will be altered. If it is, the axis of quantization needs to be changed, to a direction which is not known a priori. The H-P formalism therefore needs to be set up using an arbitrary direction for Oz and a criterion has to be introduced that, towards the end of the analysis, when the spin-wave spectrum has been obtained, will determined the optimum direction for Oz. The obvious criterion is that the ground state energy, which includes the zero-point energies of the spin-wave modes and any scalars that have been produced during the analysis, should be minimized with respect to the direction and magnitude of the magnetic field. The latter is particularly important, for antiferromagnets show a spin-flip transition as the field is increased.

This is most readily understood in terms of an elementary treatment of the susceptibility, which can be found in many standard texts (e.g., Kittel, 1986). With a magnetic field parallel to the alignment direction, the susceptibility falls linearly to zero as the temperature is lowered. With the field applied in a direction perpendicular to the alignment, the susceptibility remains constant. If, therefore, a magnetic field is applied in the parallel direction and gradually increased the magnetic energy, at low temperatures, is hardly changed. If, though, the same field is applied in the perpendicular direction the magnetic energy decreases as the strength increases. It can therefore be expected that on increasing a field in the alignment direction, a value will eventually be reached at which the total energy will be lowered if the alignment direction switches to a direction perpendicular to that of the applied field. (In spin-wave theory this instability is associated with some of the expressions for $\omega_{\mathbf{k}}^2$ becoming negative.) This phenomenon has been observed (Jacobs, 1961).

§ 9.8 Mean-Field Theory

Before leaving the topic of magnetic phase transitions mention needs to be made of a simple way of accounting for phase transitions, one which often appears to be at least as good as many of the more sophisticated treatments.

For ferromagnets it has long been known that a simple model is that in which it is supposed that a system of isolated spins are all in the same strong magnetic field, the magnitude of which is determined by the mean moments of the adjacent spins (see §5.2). The theory is readily extended to antiferromagnets by introducing an internal field that alternates in direction between the A and B sublattices. It predicts not only the phase transition at the Néel point but the low-temperature parallel and perpendicular susceptibilities and the spin-flip phenomenon. A related procedure is a modification of the H-P substitution. Instead of dropping the a^*a and b^*b operators in the expansions of the square roots they are replaced by scalars denoted by $\langle a^*a \rangle$ and $\langle b^*b \rangle$. Once the operators that describe the normal modes have been found, a^*a and b^*b can be expressed in terms of them, and their expectation values, either in the ground state or in a thermal distribution of excitations from the ground state, can be found. Setting $\langle a^*a \rangle$ and $\langle b^*b \rangle$ equal to these leads to implicit relations which when solved will give a "best estimate" for $\langle a^*a \rangle$ and $\langle b^*b \rangle$ and so a full solution of the (approximate) equations.

§ 9.9 Comments

There are many books that discuss spin-wave theory, though few actually use an anisotropic form for the exchange interactions. Also, antisymmetric exchange is usually omitted. In the context of antiferromagnetism it can lead to a slight misalignment of the A and B sublattice moments, so producing a weak form of ferromagnetism. A comprehensive treatment of a variety of applications of the H-P substitution can be found in Keffer (1966). A more recent account is to be found in a number of articles in Borovik-Romanov and Sinha (1988). This book includes an account of attempts to improve the form of the substitution. A different approach, which stresses thermal averages and correlations between spins, and which is therefore closely related to neutron scattering techniques, can be found in Lovesey (1986). For a fully comprehensive account of the use of neutron scattering, see Volumes I and II of Lovesey (1984).

§ 9.10 References

Borovik-Romanov, A.S., and Sinha, S.K. 1988. *Spin Waves and Magnetic Excitations* (North-Holland: Amsterdam), Part I.

Caspers, W.J. 1989. *Spin Systems* (World Scientific: Singapore).

Fulde, P., and Loewenhaupt, M. 1985. *Adv. Phys.* 34:589.

Fulton, S., Nagler, S.E., Needham, L.M.N., and Wanklyn, B.M. 1994. *J. Phys. Cond. Matt.* 6:6667.

Holstein, T., and Primakoff, H. 1940. *Phys. Rev.* 58:1098.

Jacobs, I.S. 1961. *J. App. Phys.* 32:61S.

Keffer, F. 1966. *Handbuch der Physik*, ed. S. Flügge (Springer-Verlag: Berlin), Band XVIII/2, ed. H.P.J. Wijn, p. 1.

Kittel, C. 1986. *Introduction to Solid State Physics* (New York: Wiley).

Lieb, E., and Mattis, D. 1962. *J. Math. Phys.* 3:749.

Lovesey, S.W. 1984. *Theory of Neutron Scattering in Condensed Matter* (Clarendon Press: Oxford).

Lovesey, S.W. 1986. *Condensed Matter Physics: Dynamic Correlations* (Benjamin/ Cummings: Menlo Park).

Martel, P., Cowley, R.A., and Stevenson, R.W.H. 1967. *J. Appl. Phys.* 39:1116.

Moriya, T. 1985. *Spin Fluctuations in Itinerant Electron Magnetism* (Springer-Verlag: Berlin).

Palmer, R.G. 1982. *Adv. Phys.* 31:669.

Rushbrooke, G.S., Baker, G.A., Jr., and Wood, P.J. 1974. *Phase Transitions and Critical Phenomena*, ed. C. Domb and M.S. Green (Academic Press: London), p. 245.

Standley, K.J. 1962. *Oxide Magnetic Materials* (Clarendon Press: Oxford).

Stevens, K.W.H. 1992. *Physica* B 176:1.

Tebble, R.S., and Craik, D.J., 1969. *Magnetic Materials* (Wiley: London).

Zeiger, H.J., and Pratt, G.W. 1973. *Magnetic Interactions in Solids* (Clarendon Press: Oxford).

10

Conductors

§ 10.1 Magnetic Oxides

Although a good deal of attention has been paid to antiferromagnetism it is not easy to think of practical applications of it. This is in marked contrast with the closely related ferrimagnetism, which is widely used. The difference arises because although both share the property of being electrical insulators the ferrimagnets have the additional property of possessing macroscopic magnetic moments. From the theoretical point of view the two are not very different, for all that would seem to be needed is to have magnetic moments on the A sites that differ from those on the B sites. In fact, the actual position is more complicated, because the technical interest in finding insulators with magnetic moments has resulted in the discovery of a wide range of magnetic crystals that have much more complicated structures than the above description would suggest. Nevertheless, it can usually be assumed that they contain either $(3d)$ or $(4f)$ ions in environments that are distorted versions of the sites in the simpler structures. Thus iron group ions are often surrounded by distorted octahedra or tetrahedra of O^{2-} ions and their properties have traditionally been described using crystal field theory, with phenomenological exchange interactions that couple them to other magnetic ions. The move from crystal field theory to its second quantized version throws a different light on this description, without meriting further discussion here. Instead, the interest will be concentrated on a phenomenon that was first recognized in a few compounds discovered in the study of ferrites. They differed from the norm in showing electrical conduction, a property that was not what was needed. Such speculation as there was assumed that it arose because the structures contained ions of two different valencies on crystallographically identical sites. There might be, for example, equivalent crystallographic sites some of which were occupied by Fe^{3+} ions and some by Fe^{2+} ions, both of which are well-known valence states of iron. It might then be possible for an electron on an Fe^{2+} ion to transfer to an adjacent Fe^{3+} ion, which simply interchanges the valencies, without an energy penalty. If further similar hops could occur this might explain the conduction. Such conductivity would thus be quite different from conventional metallic conduction, which is explained using the concept of partially filled conduction bands.

The two approaches have gradually moved closer together, for in band theory there have been a number of attempts to introduce more ionicity, such as through the Hubbard U concept. In the case of the ferrites it might seem more natural to begin with the ionic concept of localized moments and attempt to introduce some degree of delocalization. This, however, has not seemed possible within the context of crystal field theory, for it has the inbuilt restriction that every ion has a definite number of electrons. This is not a restriction in the generic Hamiltonian, and now that the spin-Hamiltonians of the ions and their mutual interactions can be derived from it there is the possibility that it will be able to cope with conduction due to electrons hopping from one magnetic ion to another. The purpose of this chapter is to explore such a possibility, not so much because it may throw light on conduction in the ferrites, but because a whole new area has recently been opened, the superconductivity found in the copper oxide compounds, the parent compounds of which are undoubtedly insulating antiferromagnets.

The first high-T_c superconductors were discovered in 1986, since when many more have been discovered and a great deal of effort has been put into elucidating and improving their properties and accounting for the origin of the superconductivity. However, there still seems (1996) to be no consensus as to how the phenomenon is to be explained, and indeed it almost seems that there are as many theories as theoreticians. The most that can therefore be said, of the theory which is to follow, is that it springs naturally out of the insulator theory, that it shows how conduction can arise, and that it has features that may turn the conductivity into superconductivity.

§ 10.2 The Prototype Copper Oxide Plane

The copper oxide superconductors have a variety of structures, none of which, so far, stray much from a basically simple structure which can be regarded as a prototype for them all. This is a copper-oxygen plane such as is shown in fig. 6.1 (p. 143) when the crosses are regarded as representing Cu^{2+} ions and the dots as O^{2-} ions. It is assumed to be insulating and composed of Cu^{2+} and O^{2-} ions. With such an ionic description it is clear that the plane is far from being electrically neutral, for it contains twice as many O^{2-} ions as Cu^{2+} ions. Indeed, it seems unlikely that such a planar arrangement could exist on its own. Nevertheless, something like it does exist in the superconductors, the difference being that the crystals consist of layers of the above type with a variety of other ions, mostly positively charged, sandwiched between them. These ions do several things: they restore the overall electrical neutrality, they result in electrons being extracted from the copper-oxygen planes, and they may play a role in destroying the translational symmetry associated with the prototype copper-oxygen plane. The basic similarity in the structures is the common possession of the copper-oxygen planes, which suggests that the understanding of the superconductivity will come from an understanding of the properties of a prototype, a copper-oxygen plane of the type shown in fig. 6.1 from which some of the electrons have been removed.

Each O^{2-} ion can be expected to have its electrons arranged so that all its occupied shells are filled. It will therefore have no magnetic moment. Cu^{2+} ions, in a square planar array, are well known in paramagnets and, using crystal field theory, they are regarded as having nine electrons in the $(3d)$ shell with the $(3d)$ "hole" being in an $(x^2 - y^2)$ orbital, where the Ox and Oy lobes are directed towards the nearest-neighbor oxygen ions. (This wave function is largely confined to the x-y plane.) Each Cu^{2+} ion then has spin $1/2$, and the expectation must be that the prototype plane will show the characteristics of a typical insulating copper antiferromagnet, due to a dominating isotropic antiferromagnetic exchange interaction between nearest-neighbor Cu^{2+} ions and a smaller anisotropic exchange interaction arising from spin-orbit interactions. (Having a diamagnetic ion between two magnetic ions usually results in enhancement of the isotropic exchange. In this structure the O^{2-} ions are between nearest-neighbor Cu^{2+} ions, which is a fairly uncommon arrangement. Usually the strongest exchange interactions are between next-nearest-neighbor Cu^{2+} ions, through an intervening O^{2-} ion.)

§ 10.3 The Conducting Plane

It seems reasonable to suppose that the plane of fig. 6.1 will become conducting if some of its electrons are removed. There are several possibilities. Some of the Cu^{2+} ions may have been converted into Cu^{3+}, and/or some of the O^{2-} may have become O^{1-}. On the whole, it seems most likely that it is the O^{2-} ions that will give up electrons, for doing so will decrease the charge imbalance in the plane. This will therefore be assumed, though it would be possible to treat the case in which the electrons are removed from the Cu^{2+} ions. (The method runs into difficulty if both oxygen and copper sites lose electrons.)

As far as possible the treatment will follow that already described for the insulators. The second quantized Hamiltonian will be regarded, at least initially, as set up using the Wannier functions for the prototype plane. The difference enters in the choice of the unperturbed Hamiltonian, for it is particularly important, for the Bloch perturbation scheme, to have an energy gap between the lowest P_0 manifold and the first excited manifold. The unperturbed Hamiltonian will, as before, be defined using projection operators, P_n, each of which is a sum of different projection operators, one for each of a set of states that have similar structures. In the insulators the similarities arose because the ions could be divided into two sets, one consisting of the diamagnetic ions and the other consisting of the magnetic ions, with the differences, which determined the number and character of the ground P_0, being due to the variety of possible spin orientations of the magnetic ions. The excited P_n's then followed by moving charges to unoccupied states.

For the copper-oxygen case the ground projection operator is more complicated and so it will be constructed in steps. The first need is to describe a set of N Cu^{2+} ions, each of which has spin $1/2$. This is most conveniently done by defining a vacuum state in which each copper ion has all its inner shells *and* the $3d$ shell filled. That is, in the vacuum state each is present as Cu^{1+}. Then the application of an operator which removes an electron of specific spin from an orbital state of $(x^2 - y^2)$ symmetry at some chosen site produces a Cu^{2+} ion at that site. Each state in the manifold that describes the totality of Cu^{2+} ions is obtained by applying a product of such operators, one for each copper site, to the vacuum state. This will produce 2^N orthogonal many-electron states, for due to the spin there are two choices for the operator at each site. The oxygens can be included, as O^{2-} ions, by defining the vacuum state so that these ions have just enough electrons to fill the appropriate shells. [The local symmetry is imposed on the Wannier functions; only those states corresponding to distorted $(1s)$, $(2s)$, and $(2p)$ ionic states are filled.] There is only one such many-electron state, so the number of states in P_0 remains unchanged. To produce an O^{1-} ion at a particular oxygen site it is necessary to remove an electron with a specific spin orientation from that site, which raises the question of which state it should be taken from. The local symmetry is different for the states of p_x, p_y, and p_z symmetry, so they are unlikely to have the same energy. The electron that is most likely to be removed is one of the two in the least bound of the three. On electrostatic grounds this is likely to be a p like orbital that is directed towards its nearest copper sites, for such an orbital results in the electron being particularly close to the electron in the copper $(x^2 - y^2)$ orbital. It will therefore be assumed that the electrons that are removed are taken from such orbitals. If r electrons are removed from O^{2-} ions and each one is left as O^{1-}, a particular configuration can be obtained by applying a product of r annihilation operators, one for each of the r oxygen sites, to the new vacuum state. As there as two such operators for each site, due to spin, this will increase the dimensions of P_0 by a factor of 2^r. In addition, removing r electrons, one at a time, from $2N$ oxygen sites can be done in $^{2N}C_r$ ways, so altogether the number of states spanned by P_0 will be $2^N \cdot 2^r \cdot {}^{2N}C_r$. All these many-electron states will be mutually orthogonal, and, in the Bloch perturbation scheme, they are all given the same energy, E_0, in the unperturbed Hamiltonian. To effect this, E_0 is chosen to be the mean of the expectation value of the actual Hamiltonian taken over all the states in P_0. Having chosen P_0, the excited states are obtained by rearranging the electrons into states that are not in P_0. They can be allocated to excited manifolds, the P_n, with $n > 0$, with each E_n being similarly defined by averaging over expectation values of states with similar characteristics. The perturbative treatment then depends on there being a finite energy separation between P_0 and the nearest excited P_n.

The whole procedure is basically the same as that used for the insulators, and it is manageable because the states in the ground manifold are all rather similar, only differing in the spatial arrangements of the O^{1-} sites and in the diversity of possible spin orientations for the Cu^{2+} and O^{1-} sites. The position would be quite different if on removing r electrons from the plane some were taken from Cu^{2+} ions and the rest were taken from O^{2-} ions. It would then be difficult, if not impossible, to decide how to group the various electronic arrangements so as to obtain an unperturbed Hamiltonian that has a highly degenerate ground manifold and a finite energy gap between the ground and the first excited manifold.

With the underlying assumption, for the Wannier functions, of periodic boundary conditions in an extended cell, the unperturbed Hamiltonian is invariant under the operations that leave the lattice invariant, as is the generic Hamiltonian. This means that the effective Hamiltonian, an operator that is bordered by the P_0 operators, will also be invariant, as well as restricted in its form. (In describing the properties of the effective Hamiltonian it is often convenient to regard it as the operator that is between the two P_0's, rather than the operator containing them. Terms in this effective Hamiltonian can be consistently dropped if they would be eliminated when the structure of the P_0's is taken into account.) The most important operators in the effective Hamiltonian can be expected to be those generated in first order of perturbation theory, on the assumption that the main effect of going to higher orders will be to produce similar operators which can be combined with those found in first order by "renormalizing" their coefficients. However, unlike the insulator examples, there is no reason to suppose that it is possible to recast these forms into angular momentum type expressions similar to those that occur in crystal field theory.

§ 10.4 Displaced Wannier Functions

It has been convenient to assume that all the electrons are in Wannier functions which have been determined by the planar structure of the prototype lattice. From a physical point of view this seems unrealistic, for on removing electrons from the O^{2-} ions in a more or less random way, not only will the translational symmetry be destroyed, but the ions can be expected to take up new equilibrium positions about which they will vibrate. From the quantum point of view the position may not be quite this awkward, because the O^{1-} ions can be expected to be interchanging electrons with adjacent O^{2-} ions, in which case no oxygen site will be either definitely O^{1-} or definitely O^{2-}. Indeed, if the interchanges take place sufficiently rapidly it would seem that the vibrations will occur about the mean nuclear positions of a square planar lattice of the type shown in fig. 6.1, with a modified lattice constant. It is this lattice that should be used in the definition of the extended lattice and so in the Wannier functions. The mean lattice constant, however, is not known.

This would make little difference if the Wannier functions formed a complete set, for the properties of a second quantized Hamiltonian do not depend on the basis. However, there is some doubt whether the concept of a complete set can be reconciled with periodic boundary conditions, as can be seen by considering a linear system, a large number, N, of equally spaced masses arranged in a circle of circumference Na. A periodic potential, with periodicity a, will provide Bloch functions, and from these Wannier functions can be defined, for any choice of a. Also, if the lattice points are given small arbitrary displacements around the circumference, it should be possible to modify the Wannier functions, by displaced Wannier transformations, so that they follow the displacements and remain orthonormal. But displacing the masses, or lattice points, in this way does not change the circumference of the circle. To do this it is necessary to change the radius of the circle and it is not obvious that there is a unitary transformation that will change a set of Wannier functions based on one choice of radius into a set based on a different radius. For a linear system that is not restricted by a periodic boundary condition (the masses lie on a straight line rather than on a circle) a change in spacing corresponds to moving the pth mass by a distance proportional to p. Such a set of displacements is not simulated by a displaced Wannier transformation, but by a scaling transformation (Stevens, 1973).

The difficulty can be partially circumvented, for the theoretician, by simply stating that periodic boundary conditions are being used and that there is an equilibrium lattice that can be used to define a good enough set of Wannier functions. Any uncertainty about the dimensions of the unit cell will then appear in the matrix elements which multiply the operators in the second quantized Hamiltonian, so all that is necessary is to regard these as parameters to be chosen a posteriori. Unfortunately, it is not usually possible to proceed to an a posteriori position without lots of approximations. Nevertheless, assuming that there is an equilibrium lattice has an important advantage, that displaced Wannier function transformations can be used.

As for the insulators, a transformation to displaced Wannier functions modifies all the matrix elements in the second quantized form, but it does not have to be accompanied by a change in the concept of what the associated annihilation and creation operators actually do, for their algebraic properties are unaltered. It is therefore convenient to regard the transformed operators as operators that produce exactly the same effects on the electronic states as they would have done had they not been transformed. The effective second quantized Hamiltonian can thus be regarded as describing a system that vibrates about an undistorted equilibrium lattice, one that has been used to define the initial Wannier functions. The physical expectation that nuclei will be moving and Wannier functions will be distorting can thus be regarded as absorbed in transformations of the matrix elements in the second quantized generic Hamiltonian. (It must, of course, include operators which describe the nuclear kinetic energy.) A result of the perturbation theory will be that the oscillations about the equilibrium lattice, induced as electrons hop from site to site, etc., will all be described in an effective Hamiltonian that acts within P_0. It will consist of combinations of "matrix elements" that multiply operator expressions in annihilation and creation operators defined using the Wannier functions of the equilibrium lattice. (The introduction of "matrix elements" needs to be explained. So far, "matrix element" has been used for a scalar quantity. The nuclear displacements are now being regarded as operators, for which no representation has yet been introduced. So the above "matrix elements" are actually operators in the Hilbert space of the nuclear variables, for the implied integrations have been made only over electronic variables.) Since the operator that displaces the Wannier functions contains the nuclear position variables, it should not be overlooked that it must also be applied to the nuclear kinetic energy expression in the generic Hamiltonian.

The dynamics associated with nuclear motion gives rise to a further consideration. In most theories of lattice dynamics the nuclear variables are regarded as contributing harmonic-oscillator-like expressions to the Hamiltonian, the phonon modes, which have what is almost a continuum of frequencies, beginning near zero and extending to a finite value. Their effect is to produce what is virtually a continuum of energy levels, above a ground level. Defining a P_0 projection operator that describes a family of degenerate states well separated from the next family presents a problem when incorporating lattice dynamics is being contemplated. The best procedure seems to be to change each P_n projection operator into a new operator, one in which the original P_n is multiplied by the unit operator in the Hilbert space of the nuclear variables, for while this greatly increases the degeneracy associated with each P_n it leaves their spacings exactly as they were before the nuclear dynamics were introduced.

The question of whether there are any manifestations of nuclear symmetry in solid state physics is usually glossed over, and it will be ignored in this chapter (though not in Chapter 11). Each nucleus will therefore be labeled by its position in the lattice. The electrons in the Wannier functions at a given site will be regarded as following the movements of the nucleus at that site. With two isotopes for copper, the expression for the nuclear kinetic energy, and its transformation when the Wannier functions are displaced, does not have the symmetry of the lattice. Apart from these terms the rest of the Hamiltonian has the symmetry of the lattice. It is therefore convenient to assume that each copper nucleus can be given the same effective mass, in which case the generic and the unperturbed Hamiltonian and the perturbation all have the symmetry of the lattice.

§ 10.5 Conductor or Insulator?

The effective Hamiltonian will contain many electronic operators which have the same form as those that occur in the magnetic insulators, particularly if the coefficients that multiply them are expanded in powers of the nuclear variables and only the leading terms are retained, which is equivalent to regarding the lattice as rigid. Indeed, the similarity becomes even more marked if all the operators that transfer electrons between O^{2-} and O^{1-} sites are dropped. The effective Hamiltonian separates into commuting parts, each of which describes an antiferromagnetic array of spin-$1/2$ Cu^{2+} ions and an array of nonmagnetic O^{2-} ions in which r of the ions have been replaced by O^{1-} ions, each of which has a spin $1/2$. The separation occurs because any state that has one arrangement of O^{1-} ions is orthogonal to any state that has a different arrangement. It then seems not unlikely that the exchange interaction between an O^{1-} ion and its nearest Cu^{2+} neighbors will be much larger than the exchange interaction between two nearest-neighbor Cu^{2+} ions, for the first pair are much closer together. If so, it can be expected that whatever the sign of the exchange between O^{1-} and Cu^{2+}, it will produce a parallel alignment of the spins of the two Cu^{2+} ions on either side of the O^{1-} ion, so locally disrupting the Cu^{2+} antiferromagnetic arrangement. (See any of the patterns in fig. 10.1, where a ferromagnetic interaction has been used.) On the lattice sites the $+$ and $-$ signs represent Cu^{2+} spin orientations and, for the most part, the oxygen sites are not apparent, though they are midway between each pair of nearest-neighbor Cu^{2+} sites, as more clearly shown in fig. 6.1. However, in each of the diagrams one O^{1-} site is shown, by a $+$ sign between two nearest-neighbor Cu^{2+} sites, which have also been given $+$ signs, to indicate that the presence of the O^{1-} ion has locally disrupted the antiferromagnetic array of the Cu^{2+} ions.

If the state $|b\rangle$ in fig. 10.1 is compared with the state $|a\rangle$, it will be noticed that the long-range part of the antiferromagnetic Cu^{2+} array is the reverse of that in $|a\rangle$ and that the two states are reflections of one another in a vertical line through the O^{1-} ion. Moving now to $|c\rangle$, it is apparent that it can be reached from $|a\rangle$ by having the O^{1-} move one step to the north-east (using a notation in which north is toward the top of the page), which is a rotated version of $|a\rangle$. As will emerge, there will certainly be spin-independent operators in the effective Hamiltonian that allow such a one-electron transfer to occur, so it can be said that $|a\rangle$ and $|c\rangle$ are connected. By similar reasoning it is obvious that $|c\rangle$ is connected to $|e\rangle$, which in turn is connected to $|g\rangle$. So there should be no difficulty in an electron hopping between these four states, though the

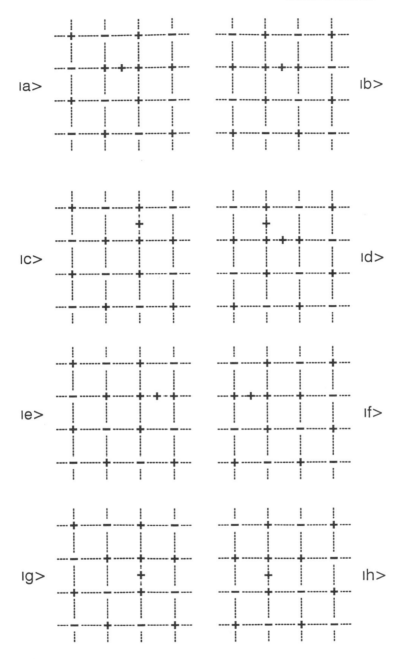

Figure 10.1. The states represented on the left have identical long-range antiferromagnetic spin structures, but different positions for a local + + + spin cluster composed of a Cu^{2+} ion, an O^{1-} ion, and a Cu^{2+} ion. The states on the right have a reversed long-range spin structure with local spin structures similar to those on the left.

somewhat awkward feature, from the point of view of conduction, is that it is not immediately obvious that it can readily hop to anywhere else. Starting afresh with $|b\rangle$, a similar argument shows that it can readily jump to $|d\rangle$, $|f\rangle$, and $|g\rangle$, but that is all. Now comes the interesting feature, that if one starts instead with $|a\rangle + |b\rangle$, $|c\rangle + |d\rangle$, $|e\rangle + |f\rangle$, and $|g\rangle + |h\rangle$, they are all connected. Further, $|f\rangle$, for example, is just a displaced version of $|a\rangle$, and reversing its long-range antiferromagnetic arrangement gives the same arrangement as that in $|b\rangle$ when it is given the same displacement. So if every $(+++)$ sequence of Cu^{2+}-O^{1-}-Cu^{2+} is associated with a long-range antiferromagnetic array of Cu^{2+} ions in which time-reversal symmetry is *not* broken, the $(+++)$ pattern should be able to move freely through the lattice. There is no obvious reason why the time-reversal symmetry of the Cu^{2+} antiferromagnetic arrangement should be broken, and as by retaining it the O^{1-} is able to move (and possibly resist the symmetry breaking) it would seem reasonable to assume that the symmetry has not been broken. [The reasoning has assumed that the Cu^{2+}-O^{1-}-Cu^{2+} has a $(+++)$ arrangement of spins. The argument is not affected if it is changed to $(---)$ or to a linear combination of the two.]

§ 10.6 Lattice Dynamics

The role of lattice dynamics is likely to be particularly complicated and, possibly, important in systems in which ions can interchange their valencies. However, even the theory of the dynamics of simple ionic structures is not particularly straightforward, so this will be briefly reviewed before going on to the more complicated case. It is convenient to do this in two stages, the first of which uses a very simple model, a chain of identical masses that are coupled by elastic interactions (see Heres, 1976, §7).

In classical mechanics there is the concept of an equilibrium arrangement, one in which the particles are at rest and equally spaced. Each mass is then given a small displacement, Q_i, in the direction of the chain, where Q_i is measured from its equilibrium position and the Hamiltonian is assumed to have the form

$$\sum_i \frac{P_i^2}{2M} + K \sum_i [Q_i - Q_{i+1}]^2, \qquad (10.1)$$

where P_i is the momentum conjugate to Q_i. The passage to quantum mechanics is made by assuming that the P's and Q's are operators that satisfy the usual commutation rules for position and momentum variables. The translational symmetry of the model is exploited by the introduction of new variables analogous to those used in the tight-binding theory of electrons in periodic lattices. With these the Hamiltonian takes the canonical form

$$\sum_k \hbar \omega_k \eta_k^* \eta_k, \qquad (10.2)$$

where η_k and η_k^* are operators that satisfy Bose commutation rules. (In many cases the details of the relation of η_k and η_k^* to the displacements is not known. That such a canonical form exists can be proved quite generally. The problem is discussed further in §10.10.) It describes a system in which the motion can be regarded as resulting from the superposition of plane waves, with the alternative description that it is an assembly of modes each of which is a harmonic oscillator with its own propagation constant k and energy level separation $\hbar\omega_k$. The reduction of the Hamiltonian to the canonical form is particularly simple when the elastic energy is taken in the form of the square of the relative displacements of nearest neighbors. [The theory can be elaborated by allowing the particles to have different masses (several ions in the unit cell) and elastic energy of a general quadratic form.] For a similar model in a three-dimensional periodic system, each mode can be characterized by a $\mathbf{k}$ vector and another index, α, which can, if desired, be incorporated into $\mathbf{k}$ by extending its definition. [α is needed because a unit cell may contain p inequivalent masses and the displacement of any nucleus has three components. With an extended cell consisting of N unit cells there will be $3pN$ degrees of freedom. $\mathbf{k}$ spans N of these, so the index α is used to distinguish between the $3p$ modes with the same $\mathbf{k}$.]

By using a simple model, the need to explain why the potential energy takes a form that is quadratic in the displacements, rather than linear plus quadratic, is avoided. It can be answered, for a system in which all the ions have filled shells, by using the Bloch perturbation scheme. For such a system P_0 will have no degeneracy from the electronic arrangement, for there is only one ground electronic state. It will, though, have considerable degeneracy because it acts in the Hilbert space of the nuclear variables. All the electronic operators in the effective Hamiltonian reduce to multiples of the unit operator in the electronic part of the Hilbert space, with the factors that multiply them being complicated expressions in matrix elements of the generic Hamiltonian.

To obtain the elastic energy in something like a quadratic form it is necessary to expand the matrix elements in powers of the nuclear variables. (Using displaced Wannier functions it is easy to overlook terms that are linear in the momenta arising from the application of the transformation to the operators describing the nuclear kinetic energy.) Interest then becomes focused on whether there are any terms that are linear in the displacements, for they can be "eliminated," with the appearance of a negative energy, by displaced oscillator transformations [see §7.5 and, particularly, eqn. (7.9)]. (The assumption that such a transformation amounts to choosing a new equilibrium position from which to measure the displacements is not always correct, for it may be necessary to combine it with a transformation that scales the size of the unit cell if the Wannier functions have been defined for a unit cell with the "wrong" dimensions.) The only other important requirements are that any terms higher than quadratic can be neglected and that the quadratic form is positive definite, which is necessary for the lattice to be stable under all small displacements.

The problem of how to choose the dimensions of the equilibrium lattice is therefore not all that simple even when the electronic ground state is a singlet, and it is unlikely to be any simpler when the electronic ground state is degenerate.

§ 10.7 The Effective Hamiltonian

The bordering P_0 operators for the Cu^{2+}-O^{2-} lattice when electrons have been removed will limit the electronic operators that can appear in the effective Hamiltonian, but not sufficiently that it is realistic to list them all. The chosen course will therefore be to devote most attention to operators that have not occurred with the insulator examples and, from these, pick the ones that look particularly interesting. The sublattice of Cu^{2+} ions, each with a spin of $1/2$, is so similar to the lattices occurring in magnetic insulators that it can be expected that the expression found using perturbation theory will contain operator forms like those that in the insulators describe antiferromagnetic exchangelike interactions between spins of $1/2$. There will be some additions, though, because the oxygen ion between each pair of nearest-neighbor Cu^{2+} ions is not necessarily present as an O^{2-} ion. Even with a rigid lattice it can be expected that if there is just one parameter, J, to describe the exchange interaction between two adjacent Cu^{2+} ions when the intervening ion is O^{2-}, there will be an extra interaction, or modified J, when the ionicity of the intervening oxygen changes. That is, J, the exchange interaction between two adjacent Cu^{2+} can be expected to change to $J + J_1(p_+ p_+^* + p_- p_-^*)$, where the operator multiplying J_1 is zero if the intervening oxygen site is occupied by an O^{2-} ion and nonzero if the ion is present as O^{1-}. The p^*, p are the operators that create and annihilate electrons in the correctly directed p-like state on the intervening oxygen ion. The bordering P_0 operators eliminate the possibility that unionized oxygen is present. (All operators should, in principle, have something to indicate the sites at which they operate. In practice these can often be omitted, without producing a loss of clarity. In the above example, the use of p^* and p indicates that an oxygen site is being considered, and the particular p-like orbital that is of relevance can be deduced, for unless it is the orbital directed towards the adjacent copper ions it will be eliminated by the bordering P_0 operators.) J and J_1 will be complicated expressions in matrix elements which, when the lattice is dynamic, will be dependent on nuclear variables.

There will be a similar expression to describe the exchange interaction between a Cu^{2+} ion and an adjacent oxygen ion when the oxygen ion is present as O^{1-}. This can be expected to involve the operator $p_+^* p_+ + p_-^* p_-$, which is effectively zero when applied to O^{2-} and unity when applied to O^{1-}. Again, the bordering P_0's will ensure that all other valence states of oxygen play no part.

The above operators are examples that can be regarded as constructed of number operators. They will be denoted by n, with subscripts as necessary. Setting $p_+^* p_+ = n_+$ and $p_-^* p_- = n_-$, the operator $(n_+ + n_+ - 2n_+n_-)$ vanishes when n_+ and n_- are both equal to unity and when they both equal to zero. However, when one is unity and the other is zero, the combination is equal to unity. So to ensure that an operator is only effective when it acts on an oxygen ion that is present as O^{1-}, all that is necessary is to premultiply it by $(n_+ + n_+ - 2n_+n_-)$.

§ 10.8 Removing the P_0 Bordering Operators

In the effective Hamiltonian obtained from the full perturbation treatment, the operators, with their P_0 borders, that involve the creation and annihilation operators of electrons on copper ions, the d and d^* operators, ensure that each copper ion is present as Cu^{2+}. An equivalent description can be obtained by replacing the bordering P_0 operators by operators that are equivalent to unity if the copper ions are present as Cu^{2+} and zero for the other states of ionization. The required operator is a product of $(n_+ + n_- - 2n_+n_-)$ operators, one for each copper site, the n's being copper number operators. Similarly, for each oxygen site the required operator is a product of $(n_+ + n_- - n_+n_-)$ operators, where the n's are now oxygen number operators. $(n_+ + n_- - n_+n_-)$ is equivalent to unity for O^{2-} and O^{1-} and zero for unionized oxygen.

The second quantized formalism, combined with the Bloch perturbation scheme, gives an expression that is bordered by P_0 operators. We now have a technique for removing the bordering P_0 operators. So rather as in spin-Hamiltonian theory, where it is usually possible to guess which angular momentum operators will occur, there is now the prospect of being able to guess what combinations of electronic creation and annihilation operators will be left. Each such combination will be multiplied by a renormalized coefficient, a complicated expression in matrix elements, each of which is likely to contain the nuclear operators as parameters. It is then possible to consider expanding these coefficients in powers of the nuclear operators. (No attempt will be made to express the effective Hamiltonian as a first quantized spin-Hamiltonian, for this does not appear to be possible. It will be retained as a second quantized expression.)

The leading terms in the expansion will be of two forms. They will either contain nuclear operators and no electronic ones, or vice versa. Indeed, if these were the only operators the dynamical motion of the nuclei would be separated from that of the electrons and the whole could be given a physical description. The various ions would appear to be existing in a rigid regular lattice that has its own, quite separate, vibrational spectrum. All the copper sites would appear to be occupied by Cu^{2+} ions, each having spin $1/2$, forming an antiferromagnetic background lattice which would be disturbed in the vicinity of any O^{1-} site. With the assumption that the antiferromagnetic array does not violate time-reversal symmetry, it is possible for electrons to jump between adjacent O^{2-} and O^{1-} sites, so allowing electrical conduction to occur. There would, however, be no mechanism by which the nuclear motion could exchange energy with any moving electrons, so there would be nothing to impede the electron flow and it would seem that the system is a perfect conductor. The last observation shows that there is a weakness in assuming that only the leading terms in the expansion need be retained. Indeed dropping the higher terms is a very poor approximation, for the vibrational spectrum associated with the nuclear motion can be expected to have harmonic-oscillator-like modes which have frequencies which extend upwards from close to zero. So even a slight interaction between the electronic and nuclear motions is likely to allow an interchange of energy between the two, however small it is. This will reduce the perfect conduction to normal conduction.

Nevertheless, superconductivity is not an unknown phenomenon, so there must be some way in which it can occur even when electrons are coupled to lattice vibrations that are much like those in any lattice. This, incidentally, raises the question of how it is possible to ignore couplings to lattice vibrations and produce a viable theory of superconductivity. Fortunately, a way of dealing with lattice vibrations was demonstrated, in the Bardeen-Cooper-Schrieffer (BCS) theory of the superconductivity found in many normal metals and alloys at low enough temperatures. Among its features it makes use of an important result, that any interaction that is linear in nuclear displacements can be largely "eliminated" by a displaced oscillator type of transformation. The "elimination" is not complete, but this is not serious, for the important feature is that the disappearance of most of the interaction is accompanied by the appearance of another interaction which does not involve the nuclear variables and which couples the electrons together in a novel way. It is this new form of interaction that, ultimately, accounts for the superconductivity. (It is easier to interpret the algebra physically rather than accurately. The analysis makes a good deal of use of unitary transformations, with the result that operators do not change their appearance. Nevertheless, their meanings change and, for example, the so-called normal modes of the lattice after a transformation are quite different from those before it: a simple lattice displacement can be changed into something that involves a rearrangement of electrons. (It is often stated that the vibrational modes have become "dressed" with electrons and vice versa.) The BCS theory then went on to show that the electronic part of the decoupled Hamiltonian has an isolated ground state with a gap to the lowest state of a continuum of electronic levels. It was the demonstration of the finite gap that led to the explanation of the superconductivity. (This description has made a number of simplifications. There is not just a single isolated low-lying state, but a continuum, each of which has its own total momentum. The concept of a ground state and a gap to an excited band of levels is associated with each value of momentum. Also, the transformation does not completely decouple the phonon modes from the electronic system. Rather, it much reduces their effect. A further reduction then occurs, due to the remarkable structures of the many-electron states on opposite sides of each gap.)

Knowing that the copper oxide structures do show superconductivity, and with an effective Hamiltonian derived from a generic Hamiltonian, it would seem of interest to go on and examine the operators that are linear in lattice displacements in much the same way as similar operators were examined in BCS theory. Before doing so, though, it is necessary to know more about what this was. That is the purpose of the next section.

§ 10.9 Negative Off-Diagonal Matrix Elements

The underlying idea can be introduced by considering a rather special Hamiltonian, one which, when written as a matrix in a special representation, has negative signs for *all* its off-diagonal matrix elements. This entails choosing "the right basis set." There is nothing unusual in having some off-diagonal elements negative, and indeed any off-diagonal matrix element, $\langle a|V|b \rangle$, can be arranged to have any sign, or phase, by changing the phase of either $|a\rangle$ or $|b\rangle$. Similarly, the members of any pair, such as $\langle a|V|b \rangle$ and $\langle b|V|c \rangle$, can be arranged to have arbitrary signs and phases by changing the phases of $|a\rangle$ and $|c\rangle$. The particular interest is in the possibility that there may be matrix elements of a cyclic nature, $\langle a|V|b \rangle$, $\langle b|V|c \rangle$, and $\langle c|V|a \rangle$, which can all be arranged to have negative signs. Giving $|a\rangle$, $|b\rangle$, and $|c\rangle$, respectively, phase changes of $\exp(i\alpha)$, $\exp(i\beta)$, and $\exp(i\gamma)$ results in the matrix elements being multiplied, respectively, by $\exp i[\beta - \alpha]$, $\exp i[-\beta + \gamma]$, and $\exp i[-\gamma + \beta]$. The product of the three phase changes is unity, so it follows that the three matrix elements cannot all be arranged to have negative signs unless the product of their initial phases is already negative. So sets of three related matrix elements each with a negative sign are a special feature.

The next step it to see why this can be important. In setting up any matrix representation of a Hamiltonian the sum of its diagonal elements will be an invariant which, in general, will be nonzero. It is then usual to subtract the mean of the sum from each diagonal element, for it describes an energy displacement that is common to all the eigenstates and so is of no interest. Assuming this has been done, the sum of the diagonal elements is then zero. It is next assumed that the Hamiltonian is such that there is a basis that makes all the off-diagonal elements negative. The states in this basis will be denoted by $|n\rangle$, where n takes integer values from 1 to N. The next step is to consider the expectation value of the Hamiltonian when it is taken over the state

$$c_1|1\rangle + c_2|2\rangle + c_3|3\rangle + \cdots, \tag{10.3}$$

in which all the c_n's are chosen to be *positive*. It is likely to have a particularly low value. The contribution from the diagonal elements is

$$\sum_n c_n^2 \langle n|H|n\rangle, \tag{10.4}$$

where

$$\sum_n \langle n|H|n\rangle = 0,$$

and

$$\sum_n c_n^2 = 1,$$

and that from the off-diagonal elements is

$$\sum_{n \neq m} c_n c_m \langle n|H|m\rangle. \tag{10.5}$$

If N is large the matrix is particularly rich in off-diagonal matrix elements, and they can be expected to produce a far greater number of contributions, (10.5), to the expectation value than (10.4). Also, (10.5) is a summation in which the individual contributions are negative, whereas those in (10.4) will be a mixture of terms with both signs. It can therefore be anticipated that the overall contribution from (10.5) will far outweigh that from (10.4), with the result that the state in (10.3) will have a particularly low expectation value.

Using the variational principle, a better approximation to the energy of the ground eigenstate will be obtained by varying the c_n coefficients, keeping them positive, so that the expectation value is minimized. All the excited eigenstates are orthogonal to it, and to achieve this the first excited state, in particular, must have at least one coefficient that is negative. The expression for its energy will then contain at least $(N - 1)$ positive contributions from off-diagonal matrix elements. Indeed, if the c_n's in the ground state are of approximately the same order of magnitude, (e.g., $1/\sqrt{N}$), there will need to be far more than one c_n with a negative value to produce a state orthogonal to the ground state. It can therefore be expected that the large number of positive contributions will result in a gap of order unity, rather than of $1/\sqrt{N}$, to the first excited state.

An example of how a finite gap can arise is usually included in any account of the BCS theory of superconductivity, with the slight difference from the above that all the off-diagonal elements are assumed to be equal either to $-V$, where V is positive, or to zero. It is then shown, formally, that in the ground state each c_n is positive and that the energy gap, found using a variational procedure, is finite as N tends to infinity. A particularly interesting result is that the expression for the energy gap, $2\omega_c \exp(-1/NV)$, where ω_c is positive, has a form that is unlikely to emerge from any perturbation method. (It is not possible to expand the exponential in powers of V.) Apart from the ground state the structures of the other eigenstates are difficult to picture in detail, though one thing seems fairly certain, that all the eigenstates will be strongly orthogonal to one another. Dealing with the case of negative off-diagonal elements with more or less random magnitudes presents an even more difficult problem, though the implication is clear that a finite gap between the ground and first excited state is to be expected when there are enough negative off-diagonal matrix elements.

BCS theory relies on showing that there is an effective Hamiltonian which contains many second quantized operators that have negative coefficients multiplying special forms with the structure $c_{1+}^* c_{2-}^* c_{3-} c_{4+}$, where 1, 2, 3, and 4 label different orbital states. The forms are related to one another by what can be described as "triangular arrangements." The operator pair $c_{1+}^* c_{2-}^*$ creates two electrons with opposite spins in different orbital states, so creating what is called a Cooper pair, and the combination $c_{2-} c_{1+}$ annihilates them. So reading from right to left the first pair in the combination $c_{1+}^* c_{2-}^* c_{3-} c_{4+}$ annihilates a $(3, 4)$ pair, if it is present, and the second pair replaces it by a $(1, 2)$ pair. The initial and final states are mutually orthogonal however many other pairs are initially present, so the effect of such an operator is to couple pairs of many-electron states by matrix elements which are negative. The triangular pattern is completed if the Hamiltonian contains $c_{1+}^* c_{2-}^* c_{3-} c_{4+}$, $c_{4+}^* c_{3-}^* c_{5-} c_{6+}$, and $c_{6+}^* c_{5-}^* c_{2-} c_{1+}$, all with negative or zero coefficients. (It is convenient to include the possibility of zero.) The various related triplets of states can be represented as in fig. 10.2. Each vertex represents a particular pair and each line represents a negative or zero matrix element which connects it to another pair. So the various triangles represent connected triplets of pairs. An isolated ground state emerges when the mesh consists of many linked triangles.

For an N-electron system the number of possible pairs, vertices, is at most $N(N-1)/2$. So if N is large the number of vertices in the mesh will be far smaller than the number of lines in it, in which case there should be no difficulty in selecting $N/2$ of the vertices, and acting on the vacuum state, which contains no electrons, by the product of the $N/2$ Cooper pairs of creation operators for that selection. An orthogonal many-electron state can then be obtained by substituting one of the vertices that has not been used. The matrix element coupling the two many-electron states will be off diagonal in this basis, and negative. Continuing this procedure will lead to the formation of a large number of orthonormal states, all of which are coupled by matrix elements which are either negative or zero (zero occurs between states that differ in two or more Cooper pairs). It is the existence of a large number of negative matrix elements

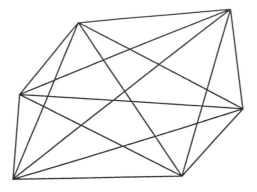

Figure 10.2. If every vertex in a pattern of dots is coupled to every other vertex, the number of lines increases rapidly with the number of vertices. N vertices give $N(N-1)/2$ lines.

multiplying operators of the form $c_{1+}^* c_{2-}^* c_{3-} c_{4+}$ that makes it possible to select the appropriate basis in which to set up the Hamiltonian and show that an isolated ground state can be expected.

The usual BCS Hamiltonian has translational symmetry, and there are separate meshes for each value of the total momentum. Each Cooper pair is associated with a specific total momentum, so the destruction of one pair, followed by the creation of another, preserves the overall momentum. There is thus a different mesh for each value of the total momentum. There may also be several meshes for the same total momentum.

The basis states used in the variational treatment of the Hamiltonian do not form a complete set, so the above reasoning needs extending. It can be regarded as a technique for producing orthogonal combinations of the kets from $|1\rangle$ to $|N\rangle$, one of which is the optimized form of that given in (10.3). This is the one that minimizes the expectation value of the Hamiltonian. The others will all be orthogonal combinations of states constructed using the Cooper pairs. To complete the basis there will need to be states that are not entirely composed of these, and the Hamiltonian will need to be diagonalized within this complete basis. This is a tall order. When and if it is ever done it would be disastrous if the eigenvalue of the first excited state was found to be above the eigenvalue of the ground state by only an infinitesimal amount. (The assumption is that, on going from the restricted basis to the full basis, the finite gap does not disappear.)

§ 10.10 The Electron-Lattice Interaction

The electron-lattice interaction is found by expanding all the coefficients in the effective Hamiltonian, so interest will now be focused on those operators that are linear and quadratic in the operators describing the nuclear positions. Some of these will just be scalar multiples of the Q's. There should, however, be no such linear terms if the underlying lattice has been chosen to be the equilibrium lattice, so they will simply be neglected. Otherwise, it will be necessary to correct, for the actual equilibrium lattice, all the analysis that has gone before, which seems an unnecessary complication when no coefficients are being given explicitly.

There is, however, a complication with the quadratic terms, which are usually written in a canonical form. It has so far been convenient to developed the theory for a model system, a two-dimensional lattice, which is a convenience that does not reflect the experimental arrangement, for any two-dimensional lattice needs to be supported. The lattice vibrations will not then be confined to a two-dimensional layer. Indeed, it is probable that the vibrational motion of the film will be dominated by the properties of the support. Fortunately, changing to a supported layer makes little difference to the formal algebra, for any quadratic potential energy expression can be brought into a canonical form. When the translational symmetry for displacements parallel to the copper oxide plane is preserved, as will be assumed, the canonical form will be

$$\sum_{\mathbf{k},\alpha} \omega_{\mathbf{k},\alpha}^2 Q_{\mathbf{k},\alpha} Q_{\mathbf{k},\alpha}^*, \tag{10.6}$$

where α is used to distinguish between plane waves of the same $\mathbf{k}$. ($Q_{\mathbf{k}\alpha}$ is a linear superposition, with complex coefficients, of the Q's of the individual nuclei. Each local Q is real and the definition results in $Q_{\mathbf{k},\alpha}$ being equal to $Q_{-\mathbf{k},\alpha}^*$. The $\omega_{\mathbf{k},\alpha}$'s are real and independent of the sign of $\mathbf{k}$ (see Ziman, 1964; Heres, 1976). Any local displacement, Q_i, can, in principle, be written as a linear sum of the $Q_{\mathbf{k},\alpha}$'s, and vice versa. An important relation is that the projection of a local Q in the ith unit cell onto a $Q_{\mathbf{k}\alpha}$ is related to the projection of a symmetry-related Q in the jth unit cell by a phase factor of the form $\exp[i\mathbf{k} \cdot (\mathbf{R}_j - \mathbf{R}_i)]$. (With a complicated unit cell the expressions for the superposition of local Q's which give the $Q_{\mathbf{k}\alpha}$'s are usually unknown. For a Bravais lattice the relation is

$$Q_{\mathbf{k}} = \frac{1}{\sqrt{N}} \sum_{\mathbf{k}} \exp(i\mathbf{k} \cdot \mathbf{R}_n) Q_n, \tag{10.7}$$

which shows the phase factors.)

It is useful when it comes to dealing with the expressions in the effective Hamiltonian which, after the matrix elements have been expanded in powers of the nuclear displacements, are electronic operators with coefficients which are linear in the nuclear displacements. As an example, it is convenient to consider the operator $[p_{i\gamma+}^* p_{i\gamma+} + p_{i\gamma-}^* p_{i\gamma-}]$, which cancels and then restores an electron at one of the oxygen sites in the ith unit cell (γ is introduced to distinguish between oxygen sites). No bordering P_0's are needed, for it is zero unless the site is initially either an O^{2-} or an O^{1-} ion. The part of its coefficient that is linear in the nuclear displacement operators contains all the displacements in the copper-oxygen plane, for they are all present in the displaced Wannier transformation. Projecting the displacements onto a particular $Q_{\mathbf{k}\alpha}$ introduces the phase factors which relate the displacements in one unit cell to those in another. So the projection onto a particular $Q_{\mathbf{k}\alpha}$ of all the $[p_{i\gamma+}^* p_{i\gamma+} + p_{i\gamma-}^* p_{i\gamma-}]$ operators for equivalent sites will be an expression of the form

$$Q_{\mathbf{k}\alpha} F(\mathbf{k}\alpha\gamma) \sum_i \exp[-i\mathbf{k} \cdot \mathbf{R}_i][p_{i\gamma+}^* p_{i\gamma+} + p_{i\gamma-}^* p_{i\gamma-}], \qquad (10.8)$$

where $F(\mathbf{k}\alpha\gamma)$ is a factor that contains a good deal of detail, largely unknown, about the precise interaction of this particular lattice mode with an electronic operator at a typical oxygen site (i.e., details of its amplitude and polarization). The projection of the same summation on to the related $Q_{\mathbf{k}\alpha}^*$ is the complex conjugate of (10.8). The interaction can now be "eliminated" by a displaced oscillator transformation which depends on the relation $\exp(-i\lambda P/\hbar) Q \exp(i\lambda P) = Q - \lambda$, where P is the momentum conjugate to Q and λ can be any operator that commutes with P and Q. For example, $\omega^2 Q Q^* + \mu Q + \mu^* Q^*$ becomes $\omega^2 Q Q^* - |\mu|^2/\omega^2$ when λ is taken to be μ^*/ω^2. In the present example, λ^* should be chosen to be $(1/\omega_{\mathbf{k}\alpha}^2)$ times the coefficient of $Q_{\mathbf{k}\alpha}$ in (10.8). The linear term disappears and another term appears, a *negative* numerical factor, $(-1/\omega_{\mathbf{k}\alpha}^2)$, multiplying the square modulus of the coefficient of $Q_{\mathbf{k}\alpha}$ in (10.8).

A similar transformation can be used for each $\mathbf{k}\alpha$ mode, with no interferences between the various modes. So had the effective Hamiltonian contained only harmonic lattice modes with linear couplings to operators of the form $[p_{i\gamma+}^* p_{i\gamma+} + p_{i\gamma-}^* p_{i\gamma}]$ it would have been possible to produce a complete decoupling of the lattice modes from the electronic motion.

Having gone so far, the next step must be to examine the properties of

$$\left| \sum_i \exp[-i\mathbf{k} \cdot \mathbf{R}_i][p_{i\gamma+}^* p_{i\gamma+} + p_{i\gamma-}^* p_{i\gamma-}] \right|^2, \qquad (10.9)$$

which is the interesting part of (10.8), for it is already clear that after the transformation it is multiplied by a factor which is negative. It can be simplified by introducing wavelike operators to describe the electronic operators, by setting

$$p_{\mathbf{K}\gamma+} = \frac{1}{N} \sum_i \exp(i\mathbf{K} \cdot \mathbf{R}_i) p_{i\gamma+}, \qquad (10.10)$$

with a similar definition for $p_{\mathbf{K}\gamma-}$. (There is a slight complication here, for $\mathbf{k}$ has, so far, been the propagation vector for a lattice wave in a three-dimensional structure, whereas the electronic operators are defined only for points in a plane. The way to allow for this is to introduce a two-dimensional wave vector, $\mathbf{K}$, which can be regarded as the projection of a three-dimensional $\mathbf{k}$ onto the copper oxide plane. The summations then range over the N^2 values for i, the number of cells in the extended cell. The definition can be inverted to give

$$p_{i\gamma+} = \frac{1}{N}\sum_i \exp(-i\mathbf{K}\cdot\mathbf{R}_i)p_{\mathbf{K}\gamma+} \qquad (10.11)$$

and the summation in (10.9) can then be reexpressed as

$$\sum_i \exp(-i\mathbf{k}\cdot\mathbf{R}_i)[p_{i\gamma+}^* p_{i\gamma+} + p_{i\gamma-}^* p_{i\gamma-}]$$

$$= \frac{1}{N^2}\sum_{i,\mathbf{K}_1,\mathbf{K}_2} \exp[-i(\mathbf{k}-\mathbf{K}_1+\mathbf{K}_2)\cdot\mathbf{R}_i]$$

$$\times [p_{\mathbf{K}_1\gamma+}^* p_{\mathbf{K}_2\gamma+} + p_{\mathbf{K}_1\gamma-}^* p_{\mathbf{K}_2\gamma-}]. \qquad (10.12)$$

On summing over $\mathbf{R}_i$ the only terms that survive are those in which $(\mathbf{K}-\mathbf{K}_1+\mathbf{K}_2)$ is zero (mod a vector of the reciprocal lattice), when each exponential factor becomes unity. Thus (10.9) reduces to

$$\left|\sum_{\mathbf{K}_1}[p_{\mathbf{K}_1+}^* p_{(\mathbf{K}_1-\mathbf{K})+} + p_{\mathbf{K}_1-}^* p_{(\mathbf{K}_1-\mathbf{K})-}]\right|^2. \qquad (10.13)$$

The cross terms between pairs that have $+$ spins and those that have $-$ spins are particularly interesting because a product such as

$$p_{\mathbf{K}_1+}^* p_{(\mathbf{K}_1-\mathbf{K})+} p_{(\mathbf{K}_2-\mathbf{K})-}^* p_{\mathbf{K}_2-} \qquad (10.14)$$

can be arranged into the form

$$[p_{\mathbf{K}_1+}^* p_{(\mathbf{K}_2-\mathbf{K})-}^*][p_{\mathbf{K}_2-} p_{(\mathbf{K}_1-\mathbf{K})+}]. \qquad (10.15)$$

The two annihilation operators on the right will cancel a Cooper pair with momentum $(\mathbf{K}_1+\mathbf{K}_2-\mathbf{K})$, which would need to have been created by

$$p_{\mathbf{K}_2-}^* p_{(\mathbf{K}_1-\mathbf{K})+}^*, \qquad (10.16)$$

and the two creation operators on the left will then replace it by another Cooper pair:

$$p_{\mathbf{K}_1+}^* p_{(\mathbf{K}_2-\mathbf{K})-}^*, \qquad (10.17)$$

with the same total momentum. The important result is that the coefficient of the operator that does this is *negative*, due to the $(-1/\omega_{k\alpha}^2)$ multiplying factor and the square modulus of $F(\mathbf{k}\alpha\gamma)$ in the expression resulting from the displaced oscillator transformation.

For clarity, the above analysis has been carried out as if there is just one oxygen site in each unit cell. In fact, there are two so the expression in (10.8) should have been summed over two values for γ. The transformation that eliminates the linear coupling to the $\mathbf{k}\alpha$ mode would then have produced three terms, one having a factor $|F(\mathbf{k}\alpha\gamma_1)|^2$, another having a factor $|F(\mathbf{k}\alpha\gamma_2)|^2$, and a third, coming from the cross product, having a factor of the form $F^*(\mathbf{k}\alpha\gamma_1)F(\mathbf{k}\alpha\gamma_2)$. The sign of the third is uncertain.

The next step is to use the BCS procedure, which leads to a ground state with a finite gap to the first excited state on using just those operators that have been shown to occur with negative signs. Again, it can be expected that the states on either side of the gap will be strongly orthogonal.

§ 10.11 Other Interactions in the Effective Hamiltonian

So far, the displaced oscillator transformations have been used to replace operators of the form $[p_{i\gamma+}^* p_{i\gamma+} + p_{i\gamma-}^* p_{i\gamma-}]$ when they occur multiplied, linearly, by lattice displacement operators. This may be misleading, because the correct procedure is to apply all unitary operators to the whole Hamiltonian, not just to parts of it. However, the variational technique has a remarkable property. It begins by choosing a set of states, each of which is composed of many Cooper pairs so that the submatrix of the Hamiltonian has off-diagonal elements from the transformed version of $[p_{i\gamma+}^* p_{i\gamma+} + p_{i\gamma-}^* p_{i\gamma-}]$, which are either negative or zero. If one now asks what has happened to the contributions from all the other parts of the transformed Hamiltonian, it is seen that they cannot contribute to off-diagonal matrix elements unless they too manipulate Cooper pairs. The places to which they are then most likely to contribute are on the diagonal, where they are relatively innocuous, due to the zero trace, provided there are enough negative off-diagonal elements.

To examine this point in more detail it is of interest to go through a parallel analysis using some other electronic form. The one that has been used has a magnitude of 2 when the site is occupied by an O^{2-} ion and 1 when the ion is changed to O^{1-}. It can therefore be regarded as an operator that simulates the effect of a force in the lattice that depends on the charge at an oxygen site. It is not the only operator that has this property. Another is $p_{i\gamma+}^* p_{i\gamma-}^* p_{i\gamma-} p_{i\gamma+}$, which is only nonzero when the ion is present as O^{2-}. Once again, its form is such that the bordering P_0 operators can be dropped. Substituting this operator in place of $[p_{i\gamma+}^* p_{i\gamma+} + p_{i\gamma-}^* p_{i\gamma-}]$ in the analysis of §10.11 makes no difference down to equation (10.13), provided that a different $F(\mathbf{k}\alpha\gamma)$ is chosen in (10.8). But beyond that there are substantial changes, because the interaction now gives rise to expressions which contain two starred and two unstarred operators, whereas previously the operator had only one starred and one unstarred operator. The summation over i that led to (10.14) now connects four momenta, $\mathbf{K}_1$, $\mathbf{K}_2$, $\mathbf{K}_3$, and $\mathbf{K}_4$, to $\mathbf{K}$, and it is far from obvious that it will give expressions like those needed to interchange Cooper pairs. (There is also another difference, in the powers of N in the normalizing factors.)

Thus the two examples, the first with one starred and one unstarred and the second with two starred and two unstarred operators, behave quite differently when the summation equivalent to that in (10.12) is performed, which shows that it is the form of the operator rather than its physical meaning that determines what happens. It then becomes relevant to ask whether there are any operators, other than $[p^*_{i\gamma+}p_{i\gamma+} + p^*_{i\gamma-}p_{i\gamma-}]$, which have the one-starred one-unstarred form after the bordering P_0 operators have been removed. The answer is that there is one other, $[p^*_{i\gamma+}p_{i\gamma-} + p^*_{i\gamma-}p_{i\gamma+}]$. (The transfer operators of the form $p^*_{i\gamma}p_{j\gamma}$, with i different from j and with any spin combination, do not retain the one-starred one-unstarred form when the bordering P_0 operators are removed. They change to three-starred three-unstarred forms.) This latest operator commutes with $[p^*_{i\gamma+}p_{i\gamma+} + p^*_{i\gamma-}p_{i\gamma-}]$, so the elimination of expressions in which it appears multiplied by nuclear displacements can be performed without upsetting the elimination of the terms of the $[p^*_{i\gamma+}p_{i\gamma+} + p^*_{i\gamma-}p_{i\gamma-}]$ form. The new form produces the expression

$$\left| \sum_{\mathbf{K}_1} [p^*_{\mathbf{K}_1+}p_{(\mathbf{K}_1-\mathbf{K})-} + p^*_{\mathbf{K}_1-}p_{(\mathbf{K}_1-\mathbf{K})+} \right|^2 \tag{10.18}$$

in place of (10.14). Its expansion contains operator combinations that interchange Cooper pairs, an example being

$$p^*_{\mathbf{K}_1+}p^*_{(\mathbf{K}_2-\mathbf{K})-}p_{(\mathbf{K}_1-\mathbf{K})-}p_{\mathbf{K}_2+}. \tag{10.19}$$

However, obtaining it in this form has required a reordering of operators, which has resulted in the introduction of an additional negative sign. When combined with the negative sign that multiplies the square modulus positive off-diagonal matrix elements are produced, which is not what is wanted in BCS-like theory. The form that has produced this result has come from operators that reverse spins at particular sites, which are only present because of spin-orbit-type interactions in the generic Hamiltonian. Since the degeneracy normally associated with p electrons has already been lifted by the environment, it is reasonable to suppose that these positive contributions to the off-diagonal matrix elements will not be enough to reverse the negative signs produced by $[p^*_{i\gamma+}p_{i\gamma+} + p^*_{i\gamma-}p_{i\gamma-}]$ operators.

When the bordering P_0 operators are eliminated for all the other operators in the effective Hamiltonian, it seems that there are no others that will interchange Cooper pairs. They will therefore contribute only to diagonal elements of the matrix set up using the sub-basis formed from Cooper pairs.

§ 10.12 General Comments

It has not been possible to give references to most of the work that has been done along the lines described in this chapter because only parts of it, the main ideas, have been published. They can be found in Sigmund and Stevens (1987, 1988). The account given here is a more recent amplification of the same ideas. The analysis shows that if, after a sequence of displaced oscillator transformations designed to "eliminate" a selection of the operators in the effective Hamiltonian, all the remaining interactions can be neglected, the operators that remain describe an electronic system which is decoupled from the lattice and which has a pattern of eigenvalues that shows gaps similar to those found in BCS theory. This result is not enough, though, to demonstrate that the system will be a superconductor. For one thing, although the effect of the displaced oscillator transformations is to decouple some of the electronic operators from the lattice, they do not decouple them all, which leaves the question of why those remaining do not destroy the superconductivity. The most likely explanation is that all the low-lying states, which are clearly very complicated linear combinations of electronic states having many Cooper pairs coupled to states in the Hilbert space of the lattice, are so strongly orthogonal to one another that they are effectively decoupled. (Even if the electronic operators were completely decoupled from the lattice variables, the eigenstates would still be states in the combined Hilbert space, though the two could then be separated. The slightest departure from complete decoupling would prevent this.)

There are one or two ways in which the present analysis differs from that for superconducting metals. One is that in BCS theory the number of electrons is regarded as variable, whereas in the present theory the number is fixed. Whether this difference is crucial is not clear, though it would be surprising if a typical BCS sample with a fixed number of electrons could not be superconducting. It seems more likely that having a variable number is convenient rather than essential. Again, in descriptions of BCS theory it often seems to be implied that there is an isolated ground level with a finite gap to the first excited level. This is not what emerges with the present theory, nor does it seem to be a feature of the BCS theory. What the present theory indicates is that it is possible to construct a very complicated sum of many-electron states, each formed by creating a lot of Cooper pairs, all of which have the same momentum. It is easy to overlook that it may be possible to begin with an initial state that has been created by selecting a quite different set of Cooper pairs, pairs, for example, having a different value for their total momentum. It is therefore possible that there will be a large number of "isolated ground states," a contradictory statement, for the use of "ground" usually implies one state. It might be better to refer to a ground manifold, meaning a large number of low-lying levels, the states of which are so strongly orthogonal to one another and to their adjacent continuum states that they are unlikely to be coupled by any of the operators which are introduced in the theoretical analysis.

Before ending this account, it may be of interest to show how an explanation of the Josephson effect might arise from the present analysis. An obvious property of a superconductor is that it is not possible to apply a steady voltage across it, for any attempt to do so will simply result in the voltage drop occurring across the leads rather than across the superconductor. It is possible, though, to separate two superconductors by an insulating barrier and create a potential difference between them, provided that the barrier is close to being a perfect insulator. The present copper oxide theory has been presented as if the sample is of infinite extent, though by the use of periodic boundary conditions a localized description has been introduced via the Wannier functions. It seems probable that in a finite sample it will still be possible to have Wannier-like functions and that much the same theoretical treatment as has been given for the infinite sample can be given for the finite sample. If so, it then becomes possible to simulate two separated superconductors, by dropping all the operators that allow electrons to transfer across a suitably placed barrier and inserting an insulator between the two parts. In the context of the copper oxide systems this would occur in the copper-oxygen plane. The restoration of transfer operators between the insulator and the conductors on either side will then provide a model that allows conduction to occur. Unlike the standard textbook description of single-electron motion from one region to another, which matches a wave function on one side of a barrier to another on the other side, the wave functions that describe the states in the present model, when the barrier is present, will be *products* of three many-electron wave functions (expressed in terms of annihilation and creation operators acting on a vacuum state, with the assumption that the lattice modes are completely decoupled), each of which is an eigenstate in its own region. It is then possible to have a voltage across the barrier. What is easily overlooked is that the wave functions in the three regions have time-dependent phase factors of $\exp(-iEt/\hbar)$ form. With a voltage across the barrier, each component in the product of the eigenstates in the three regions has a time-dependent phase factor which changes when, or if, an electron moves across the barrier. The theory needed to cope with restoring the transfer operators will have to take the time dependences of the wave functions into account. The unperturbed Hamiltonian is likely to be chosen as that in which the transfer processes between adjacent regions have been dropped, the perturbation being the part that has been dropped. Both the unperturbed Hamiltonian and the perturbation will be time independent. The unperturbed Hamiltonian will have eigenstates and eigenvalues for each different distribution of charge between the two conductors, with different time-dependent phases. (It is assumed that the charge on the insulator does not change when electrons move from one conductor to the other.) The theory will need to use the time-dependent form of the Schrödinger equation. Without going into the details, it can be anticipated that its solutions will be linear combinations of triplets of states made up of one eigenfunction from each of the three regions. Those in the conductors will have time-dependent phases which differ by factors of the form $\exp(-ineVt/\hbar)$, where e is the electronic charge and n takes a range of integer values according to how many electrons have been transferred from one side of the barrier to the other. In superconductors it is the Cooper pairs that tunnel so n will be even. Thus the states that form the basis for the perturbation theory will have a variety of time-dependent phases which differ in steps of $\exp(-2ieVt/\hbar)$. The consequence is that the expectation values of various operators, such as the electric current, for example, will show this same time dependence, for the physical consequence of the transfer of a pair across the barrier is that the state before it goes across differs from the state after it has gone by a phase factor of $\exp(-i2eVt/\hbar)$. This accounts for the periodicity observed in the Josephson effect.

§ 10.13 References

Heres, H. 1976. *Quantum Field Theory of Solids* (North-Holland: Amsterdam).

Sigmund, E., and Stevens, K.W.H. 1987. *J. Phys. C: Solid State Phys.* 20:6025.

Sigmund, E., and Stevens, K.W.H. 1988. *Physica C*: 152:349.

Stevens, K.W.H. 1973. *J. Phys. C: Solid State Phys.* 6:2191.

Ziman, J. M. 1964. *Principles of the Theory of Solids* (University Press: Cambridge).

11

Nuclear Symmetry

§ 11.1 Introduction

The generic Hamiltonian used in the previous chapters has been symmetric in nuclear as well as electronic variables, and yet in the subsequent discussion more emphasis has been placed on the antisymmetric requirement for electrons than on any symmetry requirements on nuclei. This is typical of most of solid state theory in that the requirement that the wave functions that describe nuclear motion should not distinguish between identical nuclei is either completely ignored or, in the relatively few examples where the problem is even considered, dismissed by a statement that since the wave functions of nuclei do not overlap the symmetry requirements can be ignored. Probably the most obvious case where symmetry is ignored is lattice dynamics, where effective Hamiltonians are regularly written down in a three-dimensional version of the expression given in eqn. (10.1), a Hamiltonian in which nuclei are distinguished by labeling each one by the site to which it is nearest and allowing it to interact most strongly with its neighbors, which are also distinguished by their positions. On symmetry arguments it could be argued that each nucleus should interact symmetrically with every other nucleus, a condition which, if it were to be imposed on the equation would replace $2K Q_i Q_{i+1}$, summed over i, by $2K Q_i Q_j$ summed over all i and j, so perhaps ruining the concept of phonon modes. Since such modes clearly exist, it seems that there must be some reason why symmetry can be ignored.

Even then it must be acknowledged that there are a number of examples in which nuclear symmetry cannot be ignored, the most striking examples being furnished by the two gases, H_2 and D_2. In H_2 the protons have spins of $1/2$ so all the molecular eigenfunctions should be antisymmetric under interchanges of nuclear coordinates. The rotational part of each is characterized by a quantum number, J, so it is invariant under interchanges when J is even and reversed when J is odd, with the vibrational parts being even under all interchanges. Thus when the nuclear spin states are included, overall antisymmetry under nuclear exchanges is obtained only when even J values are associated with odd values of I, the total nuclear spin, and odd J values with even values of I. With two spins of $1/2$ there are only two possible values for I, zero and unity, and the single state in $I = 0$ is odd under interchanges while the three states in $I = 1$ are even. So even J values (para-hydrogen) are linked with $I = 0$ and odd J values (ortho-hydrogen) with $I = 1$, the ortho-para notation being introduced because it is comparatively difficult for a molecule that has $I = 0$ to change into one that has $I = 1$ (Eisberg and Resnick, 1985). A consequence is that gaseous H_2 behaves like a mixture of two different gases, one of which will slowly partially convert into the other if, for example, the temperature of an equilibrium mixture of the two is changed. (There is a similar effect in D_2, which has different symmetry requirements because the deuteron has a nuclear spin of unity. All molecular wave functions then have to be invariant under interchanges of nuclei.) In both examples the energy differences involved in changing from one total nuclear spin value to another are almost entirely due to changes in the associated rotational kinetic energies.

On decreasing the temperature each gas will first liquefy and then solidify. In both phases the diatomic nature of the molecule is preserved, and if the temperature is then kept constant the ratio of para to ortho will eventually reach an equilibrium value, to produce an example of a solid state system in which nuclear symmetry plays an important role (Van Kranendonk, 1983).

§ 11.2. Hindered Rotation

An early example of what is now known to be a widely occurring effect in solids was reported by Van Kempen et al. (1964), who had been studying the low-temperature thermal properties of a series of magnetic crystals, the hexamine nickel halides. In the chloride an anomaly was observed at about 76 K and interpreted as indicating the onset of antiferromagnetism in the lattice of Ni^{2+} ions. The explanation was later revised, as indicating the temperature at which the NH_3 groups of the hexamines change from being freely rotating to librating about fixed positions. [Each NH_3 molecule consists of an equilateral triangle of hydrogen atoms, forming the base of a pyramid, with the nitrogen atom at its apex. Each nickel ion is surrounded by six such groups, with the nitrogen atoms forming a regular octahedron with the nickel ion at its center. Above the transition each NH_3 group is rotating about an axis that runs from its apex through the center of its basal triangle and on to the nickel ion. So when the rotational motion is averaged over time the overall symmetry at the nickel ion is octahedral; see Press (1981).] Below the transition the rotation is hindered by interactions between the NH_3 groups and the symmetry of the $Ni(NH_3)_6$ complex is lowered, for it is not possible to have six fixed equilateral triangles as next nearest neighbors and preserve octahedral symmetry. That it was a phase transition to an antiferromagnetic state was ruled out when it was found that the magnetic resonance spectrum showed no evidence of a substantial change in the magnetic properties. The transition to an ordered antiferromagnetic state was subsequently found, at about 1 K. But what caused most surprise was the observation of a strong specific heat anomaly at about 0.045 K, which showed an entropy change of $6k \log 2$ per nickel ion. [Using statistical mechanics it can readily be shown that if an ion has p states approximately equally populated at a temperature T_1 and only q states approximately equally populated at T_2, where T_2 is a temperature lower than T_1, the entropy change in going from T_1 to T_2 is $k \log(p/q)$. The factor of 6 suggested that the anomaly was associated with the six NH_3 groups surrounding each Ni^{2+} ion and the $\log 2$ that the number of occupied states, which could only be associated with nuclear degeneracy, had been halved.]

The temperature of the transition was far too high to have resulted from dipolar interactions between the nuclear spins, which really only left the analogue of exchange interactions between electron spins for the three proton spins per NH_3 group. Your author suggested that the anomaly should be regarded as due to an effective interaction

$$J[\mathbf{I}_1 \cdot \mathbf{I}_2 + \mathbf{I}_2 \cdot \mathbf{I}_3 + \mathbf{I}_3 \cdot \mathbf{I}_1] \tag{11.1}$$

where each $\mathbf{I}$ represents one of the three nuclear spins of $1/2$ in an NH_3 group and J represents the magnitude of a nuclear exchange interaction. With threefold symmetry all the J's would be equal and (11.1) would have just two eigenvalues, $\pm(3/2)J$, each level being fourfold degenerate. The entropy change could then be interpreted as due to an unfamiliar concept, a Heisenberg-like exchange interaction between nuclei. [At that time, exchange interactions between electrons, though they had been fairly frequently invoked, could hardly have been said to be a familiar concept, for the theory had only recently been put on a firm foundation (Anderson, 1963). The extension to nuclei, given by Thouless (1965), was still to come.] Not surprisingly, the suggestion was treated with some reserve. Since then there has been a lot more experimental work on triangles of hydrogen atoms, particularly those in CH_3 groups, which are also pyramidal and occur in many organic molecules. There is now no doubt that the nuclear degeneracies are resolved to some extent in stationary (librating) triangles of hydrogen atoms (Horsewill, 1992; Carlile and Prager, 1993).

Most of the early investigations used nuclear magnetic resonance techniques and this continues, though they are increasingly supplemented with inelastic neutron scattering experiments. The initial studies were only partially successful in determining the energy level patterns, for the above exchange interaction commutes with the total nuclear spin, which implies that transitions between states of different total spin are forbidden unless there are lower-symmetry terms, such as dipolar interactions, in the effective Hamiltonian. So whether or not the transitions could be observed depended on the strengths of the symmetry-breaking interactions and the sensitivities of the experimental arrangements. (The splittings can also be determined from the temperature dependences of the nuclear resonance transitions, a method which also has limitations.) Later developments in NMR, particularly the use of pulse techniques, have considerably increased both the sensitivity and the complexity of the experimental techniques, and there is now a wealth of information about nuclear energy levels. At the same time, the interpretations of the experimental results have become less straightforward, and what are often called experimental results are given after there has been a degree of interpretation. [A commonly used concept is that of tunneling in a threefold potential, which can be used to derive an energy level pattern similar to that obtained using (11.1). Your author is not aware of any rigorous justification of this model, which may be contrasted with that of McHale and Stevens (1990).] Unfortunately, there is a shortage of reviews of what is an extensive and interesting field of study. I am therefore indebted to my colleague, A. R. Horsewill, for drawing my attention to a recent book, Benderskii et al. (1994), which has a chapter devoted to the topic, and to three papers, Press (1981), Horsewill (1992), and Carlile and Prager (1993), which are good sources with which to enter the literature.

§ 11.3 Solid ^{3}He

A quite different manifestation of nuclear symmetry occurs with the isotopes of helium, particularly in their condensed forms. ^{4}He, the commonest isotope, has $I = 0$ so each wave function of an assembly of ^{4}He atoms should be invariant under the interchange of any two nuclei. On the other hand ^{3}He has spin 1/2 so each wave function of an assembly of ^{3}He nuclei should reverse under a nuclear interchange. (The use of the expression "interchange" leads to some ambiguity, particularly when applied to nuclear systems. From the literature it is clear that many workers assume that it means that two nuclei have actually interchanged positions, a picture that is difficult to reconcile with the concept of identical particles, apart from the difficulty of understanding how two nuclei which are kept well apart by their surrounding clouds of electrons can change places.) In the present context, interchange is interpreted purely as an operation on a wave function: an interchange of the labels given to the nuclei in writing down the Hamiltonian and which appear in the wave functions. Both helium isotopes liquefy as the temperature is lowered, but what is unusual is that neither solidifies, except under high pressure. There are therefore two phases to be compared, the liquid phases and the solid phases. In both phases the two isotopic forms behave quite differently, which, when allowance is made for the differing masses, can only be attributed to the differing symmetry requirements imposed by the nuclear spins.

In summarizing the magnetic properties of solid ^{3}He I have to thank Professor R. Dobbs, who allowed me to see three chapters of his book, *Solid Helium*, prior to its publication. It is therefore a pleasure to refer my readers to it (Dobbs, 1993) for further information. That the solid phases of the two isotopes might show different properties is not too surprising, but what is less obvious is that there is not just one solid phase for ^{3}He but several, depending on the temperature and the external pressure. The properties are so interesting that the book has three chapters devoted to the magnetic properties of solid ^{3}He, one devoted to its paramagnetism, another to the antiferromagnetism of a bcc phase, and a third to the ferromagnetism of a hcp phase. Such a diversity of properties can probably be associated with an unusual feature of each phase. Each atom appears to have a much greater freedom of movement than is found in most solids, for even though the crystal is under external pressure the lattice constant is considerably greater than that to be expected from helium atoms in contact. It seems that each atom can move quite freely until it runs into another atom. Its motion therefore seems to be rather like that of a particle in a box, except that instead of the walls being fixed they are determined by elastic collisions with other atoms. (The temperature is so low, of the order of mK, that no appreciable interchanges of energy are to be expected during the collisions.) The motion appears to be significantly different from that of the hydrogen nuclei in hindered CH_3 groups, where the scope for the nuclear movement is much more restricted by the rigidity of the CH_3 group. Nevertheless, the postulated effective Hamiltonians are similar in character, Heisenberg exchange interactions between nuclear spins, with the difference that instead of the interactions being just between pairs of spins the concept has been extended to include quartets of spins, using the form

$$(\mathbf{I}_i \cdot \mathbf{I}_j)(\mathbf{I}_k \cdot \mathbf{I}_l) + (\mathbf{I}_i \cdot \mathbf{I}_l)(\mathbf{I}_j \cdot \mathbf{I}_k) - (\mathbf{I}_i \cdot \mathbf{I}_k)(\mathbf{I}_j \cdot \mathbf{I}_l) \qquad (11.2)$$

where, as with the two-spin interactions, it is necessary to have rules to say how the four sites for the spins are to be chosen (Roger et al., 1983). Even when the new interactions are included, with the coefficients treated as parameters, it still seems to be difficult to fit all the observations. Nor do the arguments used to produce the new forms, cycles of simultaneous movements of several spins, throw any real light on what it is that determines the signs and magnitudes of the coefficients. [The reasoning is an extension of that introduced by Thouless (1965), which in turn was based on that used by Anderson in his theory of exchange interactions between magnetic ions. In the course of the development the emphasis seems to have shifted from interchange as a concept applied to wave functions to interchange as something connected with the motion of particles.] Nevertheless, the reasoning that one atom can replace another, which replaces a third, etc., is more convincing than some that is used in methyl group tunneling, where it seems to be implied that nuclear interchanges occur without there being any associated effects on the electron clouds which keep the nuclei apart.

The above examples of how nuclear symmetry can produce interesting effects may well be regarded as esoteric when compared with lattice dynamics, but they do raise the challenging question of when and why nuclear symmetry can be ignored, for it seems fairly clear that it often can be. An obvious answer is that it should never be ignored, for in the initial Hamiltonians the nuclei are not distinguished and such a fundamental symmetry should not be lost in any manipulations. Nevertheless, it is not present in the example given in §10.6, where a standard form of lattice dynamics was described, and it was lost in the reasoning that was used in the rest of that chapter. In fact, it is not difficult to see where, for it occurred in the step of replacing the static nuclear displacements by dynamic nuclear variables in the transformation that displaced the electronic Wannier functions. The rest of this chapter will therefore be concerned with an attempt to modify the transformation so that the nuclear symmetry is not lost, and to examine some of the consequences. However, the task of doing this for a system like the defect copper oxides would seem unduly ambitious at this stage, so the analysis will be restricted to models similar to the one introduced in §10.6.

§ 11.4 The Modified Wannier Displacement Operator

In the previous chapter a transformation was used to describe the distortions of Wannier functions when they are displaced to follow nuclear movements. It now appears that the transformation is unsatisfactory because it associates a particular nucleus with the electronic Wannier functions at a specific site, so labeling and therefore distinguishing nuclei. If the same physical concept is to be retained a modified transformation is required.

In the transformation, $\exp(iS)$, which is now to be discarded, S was defined in §7.5 as

$$S(\delta) = \frac{-1}{(2\hbar)} \sum_J [(\delta_J \cdot \mathbf{p})P_J + P_J(\mathbf{p} \cdot \delta_J)]. \tag{11.3}$$

Its form was deduced by beginning with the assumption that the electronic projection and momentum operators commute, for if this is so S has the property that when it is applied to an electronic Wannier function at site J it picks out δ_J, the displacement of the nucleus at that site, and the exponential operator, $\exp(iS)$, has the effect of displacing the Wannier function by δ_J. When δ_J was later interpreted as a nuclear displacement operator, J became a label for the nucleus at that site. This is not what is now needed, for the requirement is that the operator should pick out the electronic displacement at each site whichever nucleus happens to be there. The latter can only be determined, in a second quantized formulation, by nuclear Wannier functions which are associated with that site and which happen to be occupied. As with electrons there is nothing unique about nuclear Wannier functions, for they can be defined in many ways. All that is necessary is to have a periodic potential to define a system of Bloch functions, which can then be used to define Wannier functions. In particular, it is not necessary for the set that describes the behavior of the electrons to be identical with the set that describes the nuclei. As with the electronic functions, it is possible to define projection operators for the nuclear Wannier functions for particular sites. As P_J has been used for the electronic projection operator at site J, and J has also been used to denote the nucleus there, it is now necessary to make changes in the notation. To distinguish between the projection operator for the electronic Wannier functions at a site t_C and the nuclear projection operators at t_C, the former will be written as $P(t_C)$ and the latter as $\Pi(t_C)$.

To avoid unnecessary complication it will also be assumed that the lattice is linear, so that t_C and all displacements from it, which for a three-dimensional lattice would be vectors, can be taken to be scalars, which implies that all displacements are in the direction of the chain of nuclei. (The analysis can be extended to a three-dimensional model. The notation becomes much more complicated, without adding anything of real significance.) It will also be assumed, at least to begin with, that the nuclei are identical and have $I = 1/2$, though the reader may like to keep in mind the question of what differences there would be if the nuclei were bosons, so that the nuclear annihilation and creation operators have different commutation rules.

Still assuming that all operators commute, it is instructive to examine the properties of $\exp(iT')$, where T' is defined by

$$T' = \sum_{M, t_C, i} [\Pi(t_C)(R - t_C)]_M [pP(t_C)]_i, \tag{11.4}$$

where R is an operator that denotes a nuclear position, the subscript M on $[\ldots]_M$ indicates that all the contents of $[\ldots]$ relate to nucleus M, and the subscript i on $[pP(t_C)]_i$ that all the contents of $[\ldots]$ relate to electron i. The summations over M and i ensure that T' is invariant under interchanges of nuclei and interchanges of electrons. Each many-electron many-nuclear basis state can be regarded as a product of two Slater determinants, one for the nuclei and one for the electrons. If one of the nuclear determinants is expanded, a typical term in the expansion will have each nucleus in a specific nuclear Wannier function with a specific spin. Thus each nucleus appears to be distinguished by the site it is at, and it is only when the whole expansion is examined that the nuclear indistinguishability becomes apparent. A particular term in the expansion, such as $[\chi^r(t_A)]_1 [\chi^s(t_B)]_2 \ldots$ describes a state in which nucleus 1 is in the rth Wannier function, χ^r, at site t_A, nucleus 2 is in χ^s at t_B, and so on. When T' is applied to this state, the term in the summation that acts on nucleus 1 gives zero unless $t_C = t_A$, so the summation over t_C results in the state being multiplied by $(R - t_A)$ to give

$$(R - t_A)_1 \sum_i [pP(t_A)]_i. \tag{11.5}$$

[The nuclear projection operator, $\Pi(t_A)_1$, which is present in T' has been dropped because nucleus 1 is known to be at t_A.] The operator in (11.5), when used in $\exp(iT')$, has one of the required properties, that it displaces the electron at t_A by $(R - t_A)$. Similarly, the part of T' that acts on nucleus 2 has the effect, in the exponential, of displacing the electron at t_B by $(R - t_B)$. It follows that if the various operators in T' had actually commuted, as has been assumed, $\exp(iT')$ could have been used to describe displacements in which the electrons at any given site follow the nuclear displacements at that site, even when the nuclei are not distinguished. In fact, the operators do not commute, and T' is not Hermitian. The only course seems to be to change to a related operator that is Hermitian and accept the consequences, which may in fact be advantageous because with commutation each displacement occurs with no distortion of the nuclear wave functions at any site. T' will therefore be replaced by its symmetrized form

$$T = \frac{1}{4\hbar} \sum_{M,t_C,i} [\Pi(t_C)(R-t_C)+(R-t_C)\Pi(t_C)]_M[pP(t_C)+P(t_C)p]_i. \quad (11.6)$$

§ 11.5 A Simple Example?

Having obtained an operator that seems to allow the electrons at a site to follow the motion of the nucleus at that site, without distinguishing between either electrons or nuclei, the next step is to examine its use in a simple example, and what could be simpler than solid ^{3}He, with the modification that the lattice is linear? This, however, is a system that is already known to show phenomena determined by nuclear symmetry, so in seeking a reason why nuclear symmetry can often be ignored it is perhaps not the best example, except for the counter-argument that since it is rather special it may throw light on the conditions needed to determine when symmetry matters and when it does not. As the latter argument seems the more persuasive the analysis will be given for a linear chain of ^{3}He atoms.

The first step is to set up the Wannier functions, so the linear chain will be regarded as defining a sequence of identical unit cells, within each of which there will be two sets of Wannier functions, one for the electrons and one for the nuclei. To incorporate the closed-shell structure of helium it will be assumed that there is one nucleus and two spin-paired electrons in each cell, with the latter being in a definite electronic Wannier function which is a modified form of a $(1s)$ atomic function. [There is a slight inconsistency here for the Wannier functions in a given cell are not entirely confined to that cell. The correct version of the assumption is that there is only one nucleus in the nuclear Wannier functions at a given cell and that there are two spin-paired electrons in the Wannier function which corresponds to a $(1s)$ atomic function in that cell.] There is then no flexibility in the electronic arrangement, except that which arises from distortions associated with nuclear motion. There are, however, no restrictions at this stage on the forms of the nuclear Wannier functions, though these may be needed later to distinguish between the "particle in a box" motion of helium atoms in solid helium and the oscillator type of motion expected in a more typical solid.

The Hamiltonian, in first quantization, will be assumed to be made up of the usual operators, those that describe the nuclear and electronic kinetic energies and those that describe the Coulomb energies, nuclear-nuclear, nuclear-electronic, and electronic-electronic. For convenience, all spin-dependent interactions will be omitted. The second quantized Hamiltonian is then set up using a basis set that is composed of products of electronic and nuclear states, with the result that it will contain a variety of products of creation and annihilation operators for electrons and nuclei. They can be classified by their forms. The nuclear kinetic energy will contribute a sum over one-starred one-unstarred pairs of nuclear operators which are similar in form to the one-starred one-unstarred pairs of electronic operators coming from the electronic kinetic energy. Then there will be two-starred two-unstarred nuclear operators from the nuclear mutual potential energy, two-starred two-unstarred electronic operators from the electronic mutual potential energy, and one-starred one-unstarred nuclear operators multiplied by one-starred one-unstarred electronic operators from the electron-nuclear potential energy. (The only operators in the Hamiltonian that couple electrons and nuclei are those which describe their mutual Coulomb energy, which is a two-particle interaction. From the point of view of second quantization, though, it is a one-electron one-nuclear interaction.) Each combination of operators will be multiplied by a matrix element, an integral in which the integrand is defined by an energy operator in the generic Hamiltonian and appropriate Wannier functions. These matrix elements will be scalars, unlike some of those considered in Chapter 10, where the integrands also contained nuclear displacement operators but the integrations were taken only over electronic variables.

The displacement operator, $\exp(iT)$, couples nuclear and electronic variables in such a way that if it acts on a state that is a product of an antisymmetrized electronic state and an antisymmetrized nuclear state it gives a state that is a sum of product forms rather than just a single product. It is therefore best to regard it as a transformation to be applied to the Hamiltonian, keeping the basis states unaltered rather than a transformation on the basis states keeping the Hamiltonian unaltered. A relevant question is then: Does the transformation change the basic structure of the Hamiltonian, so that, for example, a one-electron operator does not remain a one-electron operator, and so on? The answer is that, depending on the operator chosen, it can be changed to a new form and therefore the form of the basic Hamiltonian can be altered. However, as it does not seem possible to write the transformed Hamiltonian in any more exact form than $\exp(iT)H\exp(-iT)$, the only course open seems to be to expand the exponentials, in which case each term in the result, which is

$$\exp(iT)H\exp(-iT) = 1 + i[T, H] + \frac{(i)^2}{2!}[T, [T, H]] + \dots, \qquad (11.7)$$

can be considered separately, for the expansion is linear in the Hamiltonian. Each modified operator form can therefore be examined separately. This raises another question: Is it a good approximation to terminate the expansion at some low order? The answer is that it may well depend on which system is being modeled. At the moment, ^{3}He is being considered and for this it is fairly certain that the nuclei make much larger excursions than they do in most solids. It therefore seems not unlikely that the convergence will be much better when the excursions are more restricted. On the other hand, although the nuclei may move further the electronic wave functions will be little distorted during the collisions because there is so little free energy available at low temperatures. In addition, the convergence is also likely to depend on which term in the Hamiltonian is being considered. For example, the expression for the nuclear kinetic energy is already quadratic in the nuclear momentum, so going beyond the first term will produce higher orders in nuclear variables. Bearing in mind that one of the reasons for investigating nuclear symmetry is to see whether an expression similar to that used in standard lattice dynamics can be obtained without neglecting the symmetry, it is reasonable to stop at zeroth order (effectively setting $T = 0$) for this operator.

For the other terms it is useful to use the property that T can be written in the form

$$\sum_{t_C} \sum_{M,i} [V(t_C)]_M [v(t_C)]_i, \qquad (11.8)$$

where, in the summation, each term is a product of an operator, $[V(t_C)]_M$, which involves only the variables of nucleus M and another operator, $[v(t_C)]_i$, which involves only the variables of electron i. To illustrate the general procedure it is convenient to use an example, the electronic kinetic energy. In zeroth order it remains as $p^2/2m$, summed over each electron, so when its expectation value is taken over the assumed closed-shell electronic structure the result is a scalar, which can be dropped as of no particular interest. In the next order it gives

$$\left[T, \sum_i \frac{p_i^2}{2m} \right] = \sum_{t_C} \sum_{M,i} \{V(t_C)\}_M \left[v(t_C), \frac{p^2}{2m} \right]_i, \qquad (11.9)$$

where use has been made of the vanishing of the commutator of two operators that refer to different electrons. Also, the commutator of any two electronic operators that refer to the same electron can be dropped, because the restriction to doubly occupied $1s$-like atomic orbitals means that only orbitally diagonal matrix elements enter (there are no spin-dependent terms in the Hamiltonian) and the diagonal elements of a commutator vanish, as is shown by taking its complex conjugate. So (11.9) reduces to zero, which makes it necessary to go to the next order in T to find any contributions that involve nuclear displacements. Attention then becomes focused on

$$\left[T, \left[T, \sum_i \frac{p_i^2}{2m}\right]\right] = \sum_{t_C, M, i} \left[T, V(t_C)_M \left[v(t_C), \frac{p^2}{2m}\right]_i\right] \qquad (11.10)$$

or

$$\left[V(t_A)_N v(t_A)_j, V(t_C)_M \left[v(t_C), \frac{p^2}{2m}\right]_i\right], \qquad (11.11)$$

where the summation sign, which is to be taken over all variables, has been omitted. (Including the details of the summations makes the expressions even lengthier: there should be no difficulty in remembering that summation is implied, unless the opposite is explicitly stated.) (11.11) separates into two terms:

$$[V(t_A)_N, V(t_C)_M]\left[v(t_C), \frac{p^2}{2m}\right]_i v(t_A)_j \qquad (11.12)$$

and

$$V(t_A)_N V(t_C)_M \left[v(t_A), \left[v(t_C), \frac{p^2}{2m}\right]\right]_i. \qquad (11.13)$$

The expectation value, taken over the electronic part of (11.12), can be dropped unless $i = j$, for $v(t_A)$ is $pP(t_A) + P(t_A)p$ and any diagonal element is either immediately zero because of the projection operators or takes the form $\langle a|p|a \rangle$, which vanishes using time-reversal symmetry. If the expectation value, summed over all i, of

$$v(t_A)_i \left[v(t_C), \frac{p^2}{2m}\right]_i \qquad (11.14)$$

is set equal to $f(t_A - t_C)$, the expectation value of

$$\left[v(t_C), \frac{p^2}{2m}\right]_i v(t_A)_i \qquad (11.15)$$

is equal to $-f(t_A - t_C)$, so (11.10) has then been reduced to an operator that involves only one- and two-nuclear variables:

$$f(t_A - t_C)\{2V(t_A)_N V(t_C)_M - [V(t_A)_N, V(t_C)_N]\}. \qquad (11.16)$$

It is convenient to consider them separately, only before doing so it is of interest to note a sum rule, that $f(t_A - t_C)$ vanishes when summed over all t_C, because $v(t_C)$ then reduces to $2p$, and the commutator in (11.14) vanishes.

If a specific site, B, is chosen and the creation operators of the nuclear Wannier functions at it are labeled B_n^*, where n takes integer values and ranges over all the nuclear Wannier functions at that site, a pair such as $B_{1\sigma}^* B_{2\sigma}$ has a coefficient, from the one-nuclear part, of the form

$$f(t_A - t_C)\langle B_1|\{\Pi(t_A)(R - t_A) + (R - t_A)\Pi(t_A)\}$$
$$\times \{\Pi(t_C)(R - t_C) + (R - t_C)\Pi(t_C)\}|B_2\rangle \qquad (11.17)$$

plus a similar expression in which the order of the two factors in the operator product, $[\ldots][\ldots]$, has been reversed. This is to be summed over all A and C. If neither A nor C coincides with B it produces

$$f(0)\langle B_1|(R - t_A)\Pi(t_C)(R - t_A)|B_2\rangle B_{1\sigma}^* B_{2\sigma}. \qquad (11.18)$$

If, however, A coincides with B but C does not it gives

$$f(t_B - t_C)\langle B_1|(R - t_B)\Pi(t_C)(R - t_C)$$
$$+ (R - t_C)\Pi(t_C)(R - t_B)|B_2\rangle B_{1\sigma}^* B_{2\sigma}, \qquad (11.19)$$

and if A and C both coincide with B it gives

$$f(0)\langle B_1|2(R - t_B)^2 + 6(R - t_B)\Pi(t_B)(R - t_B)|B_2\rangle B_1^* B_2. \qquad (11.20)$$

Although the two-nuclear terms in (11.16) have yet to be considered, as well as the contributions from the other forms in the Hamiltonian, the expressions in (11.18) to (11.20) are sufficiently interesting that it is worth pausing to examine them. An operator such as $B_1^* B_2$ has the property that it conserves the number of nuclei at site B, which is a required feature. But beyond that such a pair of operators commutes with all operator pairs at any other site. So if (11.18) to (11.20) were the only terms in the effective Hamiltonian, it would separate into a sum of commuting expressions, one for each nuclear site. This would raise the question of whether it is then possible to write down an equivalent effective Hamiltonian that describes a sum of independent entities, each of which is described by its own effective Hamiltonian. Indeed, so attractive is this possibility that it is probably worth considering two assumptions, the first being that the contributions to the effective Hamiltonian from (11.18) and (11.19) are negligible compared with those from (11.20). (In the first two expressions the presence of the projection operators results in the states at B being coupled only to states at other sites, whereas in the third the states at B are coupled only to states at B. So the assumption would seem to be valid if each nucleus is closely confined to a specific lattice site.) The second is closely related to the first, that the physically important Wannier functions in any cell can be regarded as sufficiently close to those of a complete set that they can be regarded as a "good enough complete set." (This is clearly not true under all circumstances, for it is the totality of Wannier functions that gives the complete set. It might therefore be better to rephrase this as an assumption that at the temperatures of interest the amplitudes of the nuclear motions are so small that only the lowest Wannier functions are involved in their descriptions, and it will be of no consequence if those at any particular site are assumed to be part of a complete set, the rest of which are not involved in describing the nuclear motion at that site.) (11.20) can then be replaced by a first quantized form:

$$8f(0)Q_B^2, \tag{11.21}$$

where $(R - t_B)$, the displacement from t_B, has been replaced by Q_B, a displacement measured from t_B, $\Pi(t_B)$ has been replaced by unity, and the spin index has been omitted. When it is then recalled that the nuclear kinetic energy gives $P_B^2/2M$ for the same site, the two expressions together describe a mass M, with spin $1/2$, at site B, performing simple harmonic motion about an origin. The operators at the other sites then describe similar particles (same mass and same spin, but distinguishable) performing harmonic oscillations at the other sites. Comparing this first quantized form with that of a linear chain of masses, equation (10.1), such as is usually used as a starting point for lattice dynamics, it is tempting to identify the effective Hamiltonian that has now emerged with the single-particle part of (10.1).

It is an important feature of (10.1) that its two-particle parts completely change the dynamics, from that of a system of isolated and identical Debye oscillators to a system that has a band of phonon modes, each of which describes a wave motion. The obvious question is then: Will the two-nucleon parts of (11.10), which have so far been ignored, produce the extra interactions needed to make (11.20) even closer to (10.1)?

The two-nuclear terms are actually simpler to deal with, though there is a complication with the spin indices. In (11.16) there are two summations over lattice sites, t_A and t_C, so the first step is to introduce two specific sites which will be labeled as B and D. The operator sequences that conserve the number of nuclei at B and D can be either of the Coulomb form B^*D^*DB, in which case both B operators and both D operators have the same spin, or of the exchange form B^*D^*BD, in which case B^* and D have the same spin, as do D^* and B. Taking the Coulomb form first, the two-nuclear terms factorize and reduce to

$$8f(t_B - t_D)\langle B_1|(R - t_B)|B_2\rangle\langle D_3|(R - t_D)|D_4\rangle$$
$$\times B_{1\sigma_1}^* D_{3\sigma_2}^* D_{4\sigma_2} B_{2\sigma_1}, \tag{11.22}$$

which, on the same assumptions as made above, can be interpreted as equivalent to the first-quanized operator, $16[f(t_B - t_D) + f(t_D - t_B)]Q_BQ_D$. (An extra factor of 2 enters because an identical interaction arises when B_1 is interchanged with D_3 and B_2 is interchanged with D_4.) It describes a mutual potential energy of two particles which is a product of the amplitudes of their displacements and their separation. It does not have quite the form of the corresponding interaction in (10.1), which involves the product of the amplitudes of displacement of two adjacent particles, a product which, incidentally, has the property that its coefficient equals minus twice the coefficient of the self-potential energy of each particle. The two formulae can be brought into the same form by assuming that $f(t_B - t_D)$ is zero if B and D are more than one lattice spacing apart. The coefficient of the mutual energy then becomes $16[f(d) + f(-d)]$, where d is the lattice constant and the sum rule for $[f(t_A - t_C)]$, which then becomes $f(-t) + f(0) + f(t) = 0$, shows that the coefficient of $Q_n Q_{n+1}$ is equal to minus twice the coefficient of Q_n^2. The alternative is not to make the above assumption and simply use the sum rule to show that the total potential energy is a sum of squares of relative displacements, to obtain a more general form of (10.1). This result, though gratifying, may seem remarkable. In fact, it is to be anticipated, for the sum rule is a disguised way of expressing the translational symmetry of the Hamiltonian. So the analysis appears to support, so far, the idea that nuclear symmetry can be ignored in lattice dynamics. It should not be overlooked, though, that each distinguishable particle has a twofold spin degeneracy.

The two-nuclear operators of exchange type reduce to

$$2[f(0) + f(t_B - t_D) + f(t_D - t_B)]$$
$$\times \langle B_1|R|D_4 \rangle \langle D_3|R|B_2 \rangle B_{1\sigma_1}^* D_{3\sigma_2}^* B_{2\sigma_2} D_{4\sigma_1}, \qquad (11.23)$$

where site B is always different from site D, to preserve the occupation numbers at one per site. Its presence seems fatal to the interpretation of the previous interactions, for no way has been found allowing it to be regarded as an addition to the effective first quantized Hamiltonian consisting of harmonically coupled distinguishable particles with spin $1/2$. Since, on the other hand, the experimental evidence is that in many cases nuclear exchange interactions are very small, it is perhaps in order to neglect it unless there is some particular reason to suppose doing so is incorrect.

The treatment of the Coulomb interaction between the electrons, e^2/r_{ij}, follows much the same pattern as that for the kinetic energy, the only difference being that the expression equivalent to (11.14) is

$$\{v(t_A)_i + v(t_A)_j\}\left[v(t_C)_i + v(t_C)_j, \frac{e^2}{r_{ij}}\right]. \qquad (11.24)$$

If this is put equal to $F(t_A - t_C)$ the subsequent analysis leads to the same expressions as those for the kinetic energy with $F(t_A - t_C)$ replacing $f(t_A - t_C)$. The leading term for the Coulomb interaction between the nuclei, $Z^2 e^2/R_{IJ}$, found by putting $T = 0$, gives Coulomb and exchange interactions that are independent of nuclear displacements. As it can be anticipated that there will be a good deal of shielding of electrostatic forces, it is probable that the internuclear Coulomb interactions will be greatly reduced by the electron-nuclear and electron-electron Coulomb interactions. The remaining terms are difficult to interpret, so they will not be given explicitly.

§ 11.6 General Comments

The analysis has shown that, provided one is prepared to ignore some of the terms in the expansion of the second quantized Hamiltonian found using the $\exp(iT)$ transformation and to interpret those that remain in terms of "good enough complete sets" of Wannier functions, it is possible to arrive at an equivalent first quantized Hamiltonian of the form commonly used in the theory of lattice dynamics, one in which the particles are regarded as distinguishable. The only additional feature is that each particle has spin $1/2$. It can also be seen that in arriving at this result almost no use has been made of the commutation rules for Fermi particles, for about all that has been used is that A^*A operator pairs at one site commute with B^*B pairs at another, a result that also holds for bosons. So the equivalent assumptions for a Bose system should lead to an equivalent first quantized Hamiltonian, one in which the particles are distinguished. On the other hand, such a description is unlikely to emerge if the "good enough complete set" assumption is not correct. So it is of interest to ask where this is most likely to occur.

An obvious answer is in the hindered motion of triangles of protons, for which it would seem more appropriate to assume that each nucleus is in a specific Wannier function. (The motion of each hydrogen atom is restrained by the bonding to carbon in the methyl groups.) All operator pairs of the form $A_1^* A_2$ are likely to have zero expectation values unless both A_1 and A_2 refer to the same ground nuclear Wannier function. The Coulomb-like matrix elements will then multiply operators of the form

$$A_{\sigma_1}^* B_{\sigma_2}^* A_{\sigma_1} B_{\sigma_2}, \tag{11.25}$$

which, when summed over the σ's, give a scalar, while the exchange-type matrix elements, which multiply operators of the form

$$A_{\sigma_1}^* B_{\sigma_2}^* A_{\sigma_2} B_{\sigma_1}, \tag{11.26}$$

give a scalar plus a Heisenberg interaction between the spins at A and B when summed over the σ's. This conclusion supports the form of the interaction that is used for the hindered rotators.

It is more of a problem to know what to do about solid ^{3}He, for it is by no means clear that each nucleus can be regarded as occupying just a single orbital Wannier function. On the other hand, if the spectrum of lattice vibrations at the temperatures at which the exchange effects are observed is at all like that of standard lattice dynamics, it seems incompatible to justify it using the "good enough" approximation, for this, if is valid at all, requires the neglect of the nuclear exchange interactions. So there is a dilemma; to have any exchangelike interactions seems to require an effective Hamiltonian of second quantized form, and it is only in special cases that such a form can be written in a Heisenberg-Dirac form, which is really a first quantized form. It seems that if, in the second quantized effective Hamiltonian, the exchange interactions are negligible, it is sometimes possible to replace it by a translationally invariant first quantized form. On the other hand, if they are not negligible but the nuclear motion is restricted, then a first quantized form of H-D type may be equivalent. But for a general second quantized Hamiltonian it is not obvious how it can be replaced by a first quantized Hamiltonian that includes both a phonon spectrum and isotropic spin exchanges.

A good deal of use has been made of the displaced Wannier function transformations in this chapter and in Chapter 10. However, if the two uses are compared it can be seen that it is only in this chapter that specific use has been made of the form of the operator. In Chapter 10 the actual form is not needed in expanding the matrix elements in powers of the nuclear displacements, for the only requirement is that such an expansion exists. In the present chapter the specific form is used, which raises the question of its validity. A finite displacement δ can be regarded as a step made up of a sequence of N smaller steps each of magnitude δ/N. In the second step it would then seem that the projection operator P_J ought to be replaced by the projection operator for the Wannier functions at the point to which J has just moved, and so on for all the steps. That is, at each step each projection operator P_J ought to be replaced by its transform by the S defined using the P_J of the previous step. Letting N tend to infinity leads to a differential relation for S which remains to be solved.

In the Preface I expressed the view that this book could be regarded as an attempt to meet a challenge, that of beginning with a generic Hamiltonian and produce spin-Hamiltonian forms from it. I leave it to the reader to decide on the extent to which that goal has been achieved.

§ 11.7 References

Anderson, P.W. 1963. In *Magnetism I*, ed. Rado and Suhl (Academic Press: New York), p. 25.

Benderskii, V.A., Makarov, D. E., and Wight, C.A. 1944. *Chemical Dynamics at Low Temperature* (Wiley: New York).

Carlile, C. J., and Prager, M. 1993. *Int. Journal Mod. Phys.* B 7:3113.

Dobbs, R. 1993. *Solid Helium Three* (Oxford University Press: Oxford).

Eisberg, R., and Resnick, R. 1985. *Quantum Physics* (Wiley: New York).

Horsewill, A. J. 1992. *Spectrochimica Acta* 48A:379.

McHale, G., and Stevens, K.W.H. 1990. *J. Phys. Condens. Matter* 2:7264.

Press, W. 1981. *Springer Tracts in Modern Physics* (Springer-Verlag: Berlin), Vol. 92.

Roger, M., Hetherington, J.H., and Delrieu, J.M. 1983. *Rev. Mod. Phys.* 45:137.

Thouless, D. J. 1965. *Proc. Phys. Soc.* 86:893.

Van Kempen, H., et al. 1964. *Physica* 30:1131.

Van Kranendonk, J. 1983. *Solid Hydrogen* (Plenum Press: New York).

Index

About the Author

K. W. H. Stevens is Professor Emeritus of Theoretical Physics at the University of Nottingham, England.